To my God I never Don-
Glad those bananas didn't
repen before I got this to you.
Thanks for all your help!

Trading With

DiNapoli

Levels

The Practical Application of Fibonacci Analysis to Investment Markets

Joe DiNapoli

This publication is designed to provide accurate and authoritative information in regard to the subject matter covered. It is sold with the understanding that the authors and the publisher are not engaged in rendering legal, accounting, or other professional service. If legal advice or other expert assistance is required, the services of a competent professional person should be sought.

ISBN: 1-891159-04-6

Printed in the United States of America

Warning and Disclaimer:

In commodity trading, as in stock, and mutual fund trading, there can be no assurance of profit. Losses can and do occur. As with any investment, you should carefully consider your suitability to trade and your ability to bear the financial risk of losing your entire investment. It should not be assumed that the methods, techniques, or indicators presented in this book will be profitable or that they will not result in losses. Past results are not necessarily indicative of future results. Examples in this book are for educational purposes only. This is not a solicitation of any order to buy or sell.

The information contained herein has been obtained from sources believed to be reliable, but cannot be guaranteed as to accuracy or completeness, and is subject to change without notice. The risk of using any trading method rests with the user.

A portion of the above notice is from a Declaration of Principles jointly adopted by a Committee of the American Bar Association and a Committee of Publishers.

The NFA requires us to state that, "HYPOTHETICAL OR SIMULATED PERFORMANCE RESULTS HAVE CERTAIN INHERENT LIMITATIONS. UNLIKE AN ACTUAL PERFORMANCE RECORD, SIMULATED RESULTS DO NOT REPRESENT ACTUAL TRADING. ALSO, SINCE THE TRADES HAVE NOT ACTUALLY BEEN EXECUTED, THE RESULTS MAY HAVE UNDER-OR-OVER COMPENSATED FOR THE IMPACT, IF ANY, OF CERTAIN MARKET FACTORS, SUCH AS LACK OF LIQUIDITY. SIMULATED TRADING PROGRAMS IN GENERAL ARE ALSO SUBJECT TO THE FACT THAT THEY ARE DESIGNED WITH THE BENEFIT OF HINDSIGHT. NO REPRESENTATION IS BEING MADE THAT ANY ACCOUNT WILL OR IS LIKELY TO ACHIEVE PROFITS OR LOSSES SIMILAR TO THOSE SHOWN."

The following notice applies when the names cited below are mentioned throughout the book.
FibNodes, DiNapoli Levels, Oscillator Predictor, and D-Levels are trademarks of Coast Investment Software, Inc.
Windows is a registered trademark of Microsoft Corporation..
Aspen Graphics is a trademark of Aspen Research Group, Ltd.
TradeStation is a registered trademark of Omega Research, Inc., PowerEcitor is a trademark of Omega Research Inc.
MetaStock is a trademark of Equis International.
CQG TQ20/20 is a trademark of CQG, Inc.
Market Profile is a registered trademark of CBOT.
Advil is a registered trademark of American Home Products Corporation.
Maalox and Pampers are registered trademarks of The Procter & Gamble Company.
Grecian Formula is a registered trademark of Combe Inc.

DEDICATION

This book is dedicated to my Father and Mother, Joe and Olivia DiNapoli, without whose love, persistent guidance, and care, nothing would have been possible.

ACKNOWLEDGMENTS

The production of this book was a daunting task. Its completion would not have been possible without the help of many. My deepest appreciation and thanks to:

Pat Prichard for her forbearance, diligence, love and strength.

Lee and Dave Winfield for their generosity, their talent, and their time.

Elyce Picciotti and Steve Roehl for their many technical skills and an attitude that allowed me to utilize them.

Tim Slater, Neal Hughes, and Dian Belanger for their unselfish guidance in making this project better.

My wonderful students and clients for teaching me and encouraging me to begin this project.

See Fwd.

Dan, Hank, and Carl for adding life to these pages.

Aspen Graphics™ for their permission to reproduce charts from their excellent graphics software.

My peers, some who wish to remain anonymous, many of whom go unmentioned. Thank you for sharing your knowledge with me openly and without condition: Larry Pesavento, Jake Bernstein, Bill Williams, Steve Conlon, a Chicago floor manager whose name I have long since forgotten, but who was otherwise known as "God," and last but not least, Robert Krausz.

TABLE OF CONTENTS

SECTION 1: INTRODUCTION

CHAPTER 1

TRADING METHODS
JUDGMENTAL VS. NON-JUDGMENTAL TRADING SYSTEMS
POSITION TRADING VS. INTRADAY TRADING

CHAPTER 2

PREREQUISITES,

CHAPTER 3

THE ESSENTIAL COMPONENTS

SECTION 2: CONTEXT

CHAPTER 4:

TREND ANALYSIS

CHAPTER 7

OVERBOUGHT & OVERSOLD
OSCILLATORS: WHAT WORKS, WHAT DOESN'T,

SECTION 3: CONTEXT

CHAPTER 8

FIBONACCI ANALYSIS, BASIC

CHAPTER 9

DINAPOLI LEVELS™

CHAPTER 10

DINAPOLI LEVELS™...**159**

CHAPTER 11

TRADING WITH DINAPOLI LEVELS™...............**173**

CHAPTER 12

TYING IT TOGETHER

APPENDIX

APPENDIX A

APPENDIX B

APPENDIX C

APPENDIX D

ABOUT THE TITLE:

The term DiNapoli Levels was coined by an Australian copywriter doing pre-conference promotion work on one of my speaking tours in Asia. It seemed appropriate and the attendees thought the title gave them a good indication of what the presentation was about. I also have to thank my contemporaries John Bollinger (Bollinger Bands), Larry Williams (Williams %R), George Lane (Lane Stochastics), and numerous others for giving me the fortitude to proceed with such a title. In the final analysis, however, I have to admit that this title was given serious consideration only after a certain individual managed to call successive market highs and lows after himself and... not only survived but prospered in doing so.

PREFACE

If you want a bit of background before we get involved in all the technical matters, continue to read. *Otherwise, you can jump right to CHAPTER 1 or even CHAPTER 2, without jeopardizing your understanding of my approach to trading.*

So why this book and why now? Or more broadly put, "Why would you reveal a trading method that really works? Why not just trade it? Aren't you afraid that if too many people use it, it won't work any more?" Reasonable questions that deserve answers. The short answer is that the markets have been kind to me, affording me considerable mobility and a comfortable life style. I was also recently faced with a life-threatening medical condition. Such an event gives one cause to think. Creating this book allows me to give something back. The long answer involves a bit of history.

In 1986 I experienced a severe case of emotional and physical burnout, brought on by over-trading and lack of sleep. I squandered my health and well-being for money and accolades from my peers. I learned then that there were more important aspects to life than the next tick of the S&P. On counsel from my friend, Jake Bernstein, I took a shot at the lecture circuit. It was late in 1986 in Las Vegas at the Futures Symposium International. I was totally unprepared for the response of the attendees.

The lectures were organized in two, one-hour segments, one in the morning and the other in the afternoon. I had been advised by one of the old time pros: "Give 'em an up, down, up, close, and buy. Keep it simple," he had said. "They won't understand or appreciate anything of value." "But that's not the way I trade," was my response. "So what," he muttered. I was surprised at his degree of cynicism. When I asked Jake, the event organizer, what he thought should be covered, his response was direct and simple. "Teach what you think you should. If the attendees don't like it, it's their loss." Well, that's exactly what I did. There were about 35 people in the morning workshop. Their interest was keen, their questions were intelligent and I thoroughly enjoyed sharing my knowledge.

That afternoon I spoke again. This time, the room was filled to capacity. People commandeered chairs from the hallway and other rooms. They were sitting in the aisles, on the floor, some were on top of tables at the back of the room and there were perhaps another 50 people outside the door trying to get in. About 20 minutes into the lecture, an argument broke out between those who wanted a relatively simple question answered and those who wanted me to go on. Time was very limited, and it was all I could do to prevent a brawl.

When the workshop was over, Pat my office manager, and George Damusis my programmer, were mauled for more information. Workshop attendees wanted anything they could get. We had our end-of-day graphics software (the CIS TRADING PACKAGE) which handled the trend and oscillator aspects of the lecture, but we had almost nothing on the style of Fibonacci analysis that I had taught. Fortunately we did have some FibNode™ software manuals. These did a good job of teaching Fibonacci, DiNapoli-style, and along with a few beta copies of the software, everything was gone, and I mean gone!

The next several years were loaded with speaking engagements, TV appearances, interviews, offers to manage money, possibilities to start a newsletter, fax services, and so on. While I welcomed the success, thoroughly enjoyed the teaching, and met some outstanding people along the way, it was all becoming a little overwhelming. I was also experiencing a gnawing fear that if what I had developed was overexposed, it could affect the market, my personal trading, and of course the trading of my students. To combat this possibility, I steadfastly refused to publish a book, manage money, publish any type of newsletter, or even to advertise! On three occasions I insisted on halting lectures in which there was unauthorized video taping going on. I even shelved a full video course shot at a two-day presentation sponsored by Coast in 1990, again fearing overexposure of the material. In an effort to maintain a reasonable balance, however, I created the FIBONACCI, MONEY MANAGEMENT AND TREND ANALYSIS in home trading course. I also continued development of the FibNodes™ software, and enhanced the CIS TRADING (graphics) PACKAGE. In addition, I offered some private seminars, where the number of attendees was strictly limited.

I am going through this background history to make a couple of very important points. Unlike many of my colleagues, I believe the concern of overexposure of a trading methodology - even one that has judgment involved - is *a valid and reasonable concern*. There's also a philosophical point to be made. Any professional who tries to be all things to all people, and to produce products wherever there is demand, ultimately experiences burnout. That burnout is readily evidenced in his work. I prefer to retain my focus.

Even with limited exposure, I was witnessing an effect in the market from about mid 1987 through 1990 which I believe was directly attributable to my lectures. While this effect was muted, it was nonetheless apparent. Let's be honest, if there's something good out there, it gets around. When it gets around, we need to be watchful. While its usefulness will remain significant, it is always possible that the implementation of the strategies may become more difficult.

From about 1984 to 1987 the Fib analysis, coupled with the all-important context that I will teach in this book, was so incredibly accurate, it was scary. By late 1989, massive orders being set right on Fib retracements and objective points, with stops two or three ticks away, began to wreak havoc with the *unqualified* Fib players. While I could compensate my own techniques for what I was observing, the casual Fib player was beginning to get hurt. If a lot of traders get on to something really good, the market will see to it that the majority will *still lose*. It has to be that way in order for the market to work.[1]

Fortunately, in December of 1989, *Technical Analysis of Stocks & Commodities* magazine published an article by some professor type with a doctorate in pure mathematics. He did a "study" on the validity of Fibonacci movements in the markets. The study "proved" by irrefutable, geometric, logic to any reasonable cerebral type, that the methodology of applying Fibonacci analysis to the markets just didn't work.[2]

I found out about this "authoritative" article while speaking at an economics conference in Chicago in 1989. I was with a good friend and client, a professional floor trader, when several attendees excitedly bounded toward us. The first individual was waving a copy of *Stocks & Commodities* magazine around, all excited about this article, that "thoroughly disproved" the subject matter of my upcoming lecture.

When my client and I figured out what all the excitement was about, we simultaneously gave each other the trader's equivalent of a "high five." It's something like a war dance. The attendee was not a professional and couldn't comprehend our joy. Why would the Fibonacci Guru be happy about an article written in *Stock & Commodities* magazine, that maintained Fibonacci analysis didn't work? Of course, what we hoped as professionals, and what the newcomer couldn't fathom, was that our job as professional traders was about to become easier - hopefully a lot easier. Thanks to the difficulty the market was presenting the *casual* Fib player and that magazine article, over the next weeks and months, that's exactly what happened.

So, for me, revealing my trading methodology was a balancing act between the negative effect such exposure might have, and the many, many, benefits that ensue from being recognized as an authority in one's field of expertise. Many of you who are now simply striving for that winning trade cannot imagine the doors that expert status in trading opens for you - not only here in the US, but all over the world!

Starting around 1991, I began to shift my focus to Asia. The markets were ripping over there and I was getting constant input from clients located in Asia and elsewhere around the world that my methods were barbecuing the competition. I've always taught students to go where the profits were easiest and I had always wanted to be in Asia, so... I pulled the plug on all but the best speaking engagements in the US and set out on an exploration

---

[1] Joe DiNapoli, *FIBONACCI, MONEY MANAGEMENT AND TREND ANALYSIS in home trading course* (Coast Investment Software, Inc.).
[2] Herbert H. J. Riedel, "Do stock prices reflect Fibonacci ratios?" *Technical Analysis of Stocks & Commodities,* December 1989.

of all the wonders that collectively are Asia. As a professional, speaking in all the major centers, I was able to gain insight into the culture and to gather "competent information" on how the markets functioned in each of the countries I visited. It was a fascinating experience. Back in the US, clients were still finding their way to my door, but the numbers were manageable, and more importantly, *the effect on the markets remained muted.* Now, at this writing in 1997, things are looking pretty good for additional exposure of the material. There are a number of new Fibonacci experts, some, former students. There has been some good work done. There have also been a few, let's call them "unworkable" books on the subject. Consider this. If an "unworkable" book is written on Fibonacci analysis and people lose money attempting to employ its techniques - that's a good thing. Anything that attracts traders away from the utilization of the concept *in its proper* form is an advantage to those who are skilled in how to employ it. There will be less "destructive" activity in price areas important to us.

Everyone has his own idea on how this methodology should be applied. Everyone is doing something different with it. Plainly speaking, this has made it easier for me to trade and has made it *possible for me to write this book.* The simple truth is, the more people who teach a different or impractical use of Fibonacci analysis, the better it is for me - and for you. The idea that the whole concept will be denigrated entirely is a virtual impossibility. There are simply too many people making too much money trading it - if they know how to apply it properly.

Aside from Fibonacci work, the marketplace abounds with new techniques and methods. There are eager new traders with eager new experts to teach, and brand new methods to be taught. TradeStation® and other software packages with their blind system-building techniques, put large numbers of orders in lots more places. All this is good news. It leads to the likely conclusion that a reasonable understanding of what is contained herein will pay big dividends. But beware! *If* this book catches on in a big way, and *if* it is widely followed (unlikely since there is work involved), over time there may be some consequences. If any methodology is spread too widely spread, the market typically will allow only those who study thoroughly and pay attention to nuances, to fully recognize its true potential and promise. In that respect, this approach is no different from others.

SECTION 1

INTRODUCTION

This is a book about a comprehensive and modular trading approach that I have found to be prudent and highly effective. It is about the ***PRACTICAL*** application of Fibonacci ratios to investment markets. In order to implement these Fibonacci based strategies successfully, a considerable *foundation* and structured *context* must be put in place. The text contains 15 information-packed Chapters, a comprehensive set of Appendixes, and Reference material, as well as an orientation in the form of a Preface. *Fibonacci techniques are not taught until Chapter 8*, so that the ground work can be properly laid. If you choose to gallop ahead, it is my hope that you have already formulated a comprehensive context in which to use the powerful *leading indicator* techniques, referred to herein as DiNapoli Levels .

CHAPTER 1

TRADING METHODS

JUDGMENTAL VS. NON-JUDGMENTAL TRADING SYSTEMS
POSITION TRADING VS. INTRADAY TRADING

GENERAL DISCUSSION:

Judgmental approaches call upon the trader to make decisions within a given criteria or context, while non-judgmental systems are strictly mechanical.

The trading methodology I use involves judgment. It's the way I like to trade. I believe judgmental techniques have inherent advantages over non-judgmental techniques. Flexibility as inspired by the human mind, and the speed with which necessary adjustments can be made to respond to changing market conditions are two of the strongest reasons to trade using judgment. I know from teaching however, that many of you have preconceived notions about different approaches to trading that are inconsistent with reality. Since achievement in any field first involves some fundamental understanding, I thought it might be in your best interest to spend a little time looking at certain realities of basic trading approaches. First we'll go through a reality check, then some history, so you can see why and how I have reached certain conclusions. We'll consider judgmental vs. non-judgmental trading methods then position vs. intraday trading approaches.

REALITY CHECK:

The Beach Boys had a song many of you should remember that began, " Wouldn't it be nice ...." It extolled the virtues and fun of constant companionship, where lovers could skip through the daisies to the never-never land of untold bliss. After not so many years, some of the Beach Boys themselves had to face the realities of married life, legal hassles in divorce court being one. Some who have avoided such a fate would privately admit that their former soul mates have become little more than dents in the bed next to them.

Are things always that bad between expectation and reality? Of course not. With enough effort, we can hopefully end up somewhere in between. Likewise, such is the promise of both judgmental methods and non-judgmental trading systems. Let's look at non-judgmental trading systems first.

NON-JUDGMENTAL APPROACHES:

WOULDN'T IT BE NICE...

1. Once you have the development in place, your research and your work is over. Your trading system is fixed, stationary, and immutable. Stress is non-existent since the decision-making process is out of your hands and in the purview of a machine. Thorough and precise (hypothetical) testing techniques have left little to chance. Everything has been taken into consideration so your confidence is strong.

2. You can arrange for signals to be implemented by a hired trader or broker and thereby avoid the tedium of monitoring the markets yourself.

3. "It" - i.e. the "system," the "program," the "solution" - can generate an adequate income flow which will enable you to go to Fiji and stick your toes in the sand. Perhaps "it" can even pay the alimony and child support while you find a new soul mate to take the place of the dent in the bed.

THE REALITY...

1. The work never ends. When system historical extremes are exceeded, you're back to tweaking, testing, and massaging the parameters.

1A. In fact, you better have two independent systems, or maybe three, or four, to even out the equity swings. Oh yes, there will not only be some tweaking and massaging, it's more likely that outright replacement of a system or two will occur, as one or the other goes sour altogether.

1B. What about stress? You don't know stress until you experience pervasive impotence. You feel utterly helpless to effect results as you watch your system(s) dictate one absurd order after another. You just know that profit is going to evaporate and go to a loss. When that happens, you can't do *anything* but watch and obey the signals it generates. Hey Mack, pass the Maalox®, NOW!

1C. You learn that the $100 for slippage and commission you thought was extravagant, was in fact sorely inadequate. You forgot about limit moves, 40 tick runs without looking back, worst possible fill situations and.... The data you did the testing with was thought to be okay, but really wasn't all that good. Your confidence in your testing techniques is hitting new lows along with your account size.

2. The broker you have executing the trades seems to be missing entries on some of the biggest moves and ... *why couldn't he get that stop right!* Or, the trader you hired can't help but put his vast experience (one year) to work "improving" what you have struggled so long to perfect.

3. The only way to get enough capital to properly fund four systems over the 15 futures contracts you have found necessary to trade for adequate system diversification, is to take in, and manage money. Now you have disclosure statements, CFTC (Commodities Futures Trading Commission) oversight, a staff, and more NFA (National Futures Association) compliance issues than you ever dreamed existed. You thought a corporate tax return was hard to comply with? Now the manure is so thick, you wonder if the daisies you planted will ever have an opportunity to germinate, much less see the light of day.

WHAT ABOUT JUDGMENTAL APPROACHES?

WOULDN'T IT BE NICE..

1. You study under the best of the high-powered pros. You achieve a 90% win ratio through unexcelled market understanding.

2. You live where you want, trade when you want, and rid yourself of the employee hassles that have been bugging you for years.

3. You turn a modest amount of money, through skill and diligence, into a veritable mountain of pure financial muscle. You leave an ever-growing portion of it in a high-yielding money market account whose proceeds will allow you to zip off to Fiji when you choose to and... well, you know the rest of the story.

THE REALITY...

1. You learn from one high-powered pro, then from another high-powered pro, and although you find some real benefit here and there, you're just never able to achieve quite what you expected.

    1A. In fact, it's been years now, and after $30,000 in seminars, books, software, and trading courses, your profit is only barely able to cover your cost of overhead. You haven't been able to touch that $50,000 you're out from past trading losses!

2. If you can't find a way to get *really* profitable and hit a big home run, your savings will be gone in not so many months. You start wondering if you will become someone else's employee.

3. The stress and time demands of constantly focusing on that screen, being there for the open, and tangling with the overnight Globex session have you wondering if you'll ever get to the local beach. As for Fiji, is there a contract on them? What's the tick size and... where do Fijis trade?

Okay, things aren't quite as rosy or quite as bad as what I've outlined above, but they easily can be. In fact, they can be worse! What follows are some unqualified observations from my direct involvement: the odyssey *I have experienced*. This recantation of events isn't hypothetical, it's real time living stated here, so you can see how I arrived at certain conclusions. Then, perhaps you can better decide for yourself where you might best fit in.

SOME HISTORY:

Around 1980 I decided to investigate the futures markets. The plan was to switch from the vehicles I had been trading to what I knew was the most demanding and potentially rewarding game in town. The timing of the switch had to do with my view of my station in life. By that time I was "well heeled" enough to take what I expected would be a rough transition. I also thought my knowledge level and trading expertise were at a point where this new challenge could be properly met. I quickly realized two things. It was a good thing I had waited until I was "well heeled" and the challenge was a bit more than I had anticipated. Here are the highlights of the odyssey, how I got started in, and eventually became successful at trading futures.

After about a year of trading poorly, I managed a much sought-after meeting through a social contact with an extremely successful and reclusive CTA. This man was reputed to have made a bazzilian dollars over the past five years in agricultural futures. I wish I could describe this bizarre individual to you, but perhaps someone would recognize him, and one prerequisite for his tutelage was that I never reveal his identity. Of course, I never have.

After a few pleasantries, this "cerebral" type began our get-together with a discussion of, "What if you were a Martian and you came to earth to trade commodities." Hmm.... "You look at this action, that action, another action, and so on. Not being able to speak English, you simply watch actions. Prices fluctuate." He went on, "You talk with your Martian friends about these actions and you wonder about appropriate reactions." I looked at him as if he were Moses holding the Commandments behind his back and musing about the fit of his sandals. After an hour of this much cloaked "benefit," I was so befuddled and confused I was willing to settle for which way soybeans might be going. Hopefully that information would allow me to get back the cost of my travel to see him!

This man was the first of three mentors I was fortunate enough to have. Their kindness and willingness to share with me weighed heavily in my decision to begin teaching in 1986.

So what was all this about Martians? It took me a while to figure it out, but he held the key to the gold box and he wasn't about to share it with a stranger whose intentions, interest, and sincerity were unknown. This meeting was the first of many that stretched over a period of about three years. I learned a lot from this man who traded strictly and competently from the basis of a non-judgmental system. But, strangely enough, none of what I learned was what I was after in that first meeting. He taught me that:

1. There are no absolute heroes, only heroes for a while.

2. All non-judgmental systems eventually fail (stop making money). Your hope is to be using one while it is working.

3. Excellent information can be gleaned from a true expert once trust, eligibility, and pre-requisites in your knowledge base are established.

4. Non-judgmental systems are lucky to achieve 50% winning trades, 30% is acceptable.

5. Trading a non-judgmental system is difficult and stressful. It requires tremendous concentration, diligence, and self discipline.

6. There is an awesome level of challenge, fulfillment, and discovery in the entire trading process.

LET'S ELABORATE ON SOME OF THESE POINTS:

1. HEROES:

It turns out my friend was a hero of truly epic proportions during the market moves of the 70s. His non-judgmental, fundamentally based, mathematical system was very cleverly put together. It fell apart, however, after the inflation peak of the late 70s, and after subsequent avenues of supplies (grains) opened up overseas.

2. SYSTEM FAILURE:

With substantial personal resources, cash, and experience, he diligently proceeded with a staff and mainframes to replace what he had lost. One of the more interesting dead ends he explored was a detection of randomness, or lack thereof, to determine if a tradable trend existed. It was sort of like a Directional Movement Index gone mad. That one worked great for two years, then it stopped making money. When later it bombed out, it bombed out big. Through consultations with other traders, joint ventures and the like, he was funneled down the corridor of many of my other system-developing friends: i.e. that some form of volatility breakout system seems to best handle the test of time. As most would agree, however, these types of systems have a poor win/loss ratio and generally take significant funds to employ due to their diversification requirements. There was some additional hope gleaned from all this number crunching: i.e. some systems can last up to five or perhaps even 10 years before falling apart, and if you happen to get on to one of those early on, you can do awfully well, at least for a time.

3. APPRENTICESHIP:

My naive idea that this individual would share his earned knowledge with me on our first encounter was absurd. That he would recognize my sincerity and my obvious worthiness was also absurd. It took years of doing for him, before he shared with me. He knew I needed to be ready to hear what he had to tell me. He also knew that when he told me what he knew, *I would realize how little he knew.*

That's exactly what happened. When we reached that point, I went on. In the many encounters with bright and successful traders of fixed systems in the ensuing 16 years, I have not had reason to materially change what I had learned from my first mentor.

During the time I was working with this man, I ran into my first really good trading tip, given to me by my second mentor. He was an extremely successful *judgmental* trader who, I think out of pity, told me to study Displaced Moving Averages. Of course, after telling me to study DMAs, he had to explain what they were. By doing so, he was finally able to get me out of his presence and go back to his mainframe.

It seemed in those days the only truly successful traders I found all had mainframes and they were incredibly eccentric and reclusive. I'd spent less than 15 minutes with this person, but now I had a direction. Three years later when I compared the results of my research with his, the similarities were astonishing. It took another 15 minutes to compare our research. This was the second and final meeting I ever had with this man. Now I, too, had a profitable and reasonably consistent judgmental method to trade.

The method wasn't all that hot however, about 50% winners, and it gave back a lot. Although I had a reasonable methodology, I was totally unsatisfied. Necessity being the mother of invention led to my first important independent discovery, the "Oscillator Predictor™," a true leading indicator. With it, I was able to capture profit and avoid risky entries.

It wasn't until my third mentor told me about an Italian mathematician named Fibonacci[1], that my techniques really began to click. Number three was also eccentric. He didn't own a mainframe and was anything but reclusive. This man was certainly the most dazzling trader I had *ever* met. Definitely judgmentally-based, I actually saw him nail down highs, lows, intermediate rally highs, intermediate retracement lows. You name it! His trading style went beyond any rational expectation I had ever entertained, and he did this live time while I watched him! It wasn't until he taught me what he was doing in one short afternoon, that I discovered these techniques weren't nearly enough for consistent, profitable trading. As experience (time) robbed yet another hero from me, I saw him go from riches to rags, to insolvency, to debt. The Holy Grail sure seemed to have a lot of holes that needed mending. After many, many years of my own experience, the following hard-won conclusions make up the substance of the thread.

JUDGMENTAL TRADING:

1. The most important trading tool you have is not your computer, your data service, or your methodology. IT'S YOU! If you're not right - *you don't trade.*

> 1A. Trading breaks are essential, particularly for intraday players. Three to seven day breaks every three to six weeks is what I have determined is best for me.

> 1B. Take significant time off; about three to six months every year, if you trade intraday, at least one to three months if you trade daily-based or above.

> 1C. Four or five days a week spend at least one hour doing something you like that does not involve the markets or your computer. I like working with my hands, restoring cars or otherwise fixing or building things.

> 1D. The definition of a professional is one who makes the least mistakes, not the one who makes no mistakes. If you make one serious mistake, take yourself off trading for three days. If you make three mistakes in a short time period (over two consecutive days), take yourself off trading for three days.

---

[1] Leonardo de Pisa, was the son of Guilielmo Bonacci. In Italian, 'figlio' means 'son,' hence 'figlio Bonacci,' which was shortened through the years to Fibonacci. He was a preeminent mathematician of his time.

1E. If you violate rule 1D, give $100,000 to the person you dislike the most in the world. It may interest you to know that of the many traders I have trained, I have almost never seen anyone follow rule 1D unless they were forced to (no margin). A mistake is defined in CHAPTER 2, under "Ground Rules and Definitions." Mistakes will be cited and clarified throughout this book. It's critical that you know if and when you make a mistake .

1F. If you have not been trading for 10 days or more, do not trade size (large positions) for at least one week.

1G. Separate yourself into two halves, the trader and the manager. The trader cannot trade without the express permission of the manager. The manager watches for crucial signs of "fitness," e.g. mistakes, irritability, stress in the trader's personal life, tell-tale black shading under the eyes, excessive flatulence. Get the idea? It is the manager's job to get the trader away from the phone before the disaster occurs. Just look at the Barings fiasco if you doubt the need for competent management.

In the late 80s, a local I had trained who had achieved consistent and considerable profitability, began just as consistently, losing money. His personality seemed to be undergoing a change and it was obvious his personal life was under some stress. I first suspected drugs, but on further reflection I narrowed it down. One night when we were talking market and he was complaining bitterly about "unexplained" losses, I asked him how long it was before his new child was to be born. "Three months," he said, "and how the hell did you know?"

By this time, you should be able to see the importance of being "fit" to trade. Systems or methods allowing for discretion are dependent on the quality of that discretion both in execution and size. If the discretion goes, you can wipe out in a week what it has taken you months to accumulate.

2. There are knowledgeable, honest, and sincere traders available who like to teach and have the qualifications to make themselves understood. Seek them out, get their material if you can afford it, and befriend them. Much of what they can offer to you in casual conversation will be invaluable to your success. Someone helped them. Approach them correctly and if they can, they will help you.

3. Judgmental trading can lead to incredibly favorable win/loss ratios. Don't let them go to your head.

> 3A. If you attain dramatic profits very quickly, your ego can blow way out of proportion. Always remember *you are only one trade away from humility.*

4. During the trading decision-making time, you can afford no interruptions. None! Zero! If your office is in your home and you trade intraday, get a lock to separate yourself from anyone sharing the premises *and use it.* If this presents a problem with your wife or family, don't trade or get a different wife and family.

One of my clients, a chiropractor, who traded out of his home in northern California, was warned about this repeatedly. This individual suffered a $40,000 loss when his wife casually walked over and dropped their infant in his arms just before a crop report. Experience can be a tough teacher but to many it's the only teacher.

> 5. The speed with which you can adapt to changing market conditions is considerably faster with a judgmental rather than a non-judgmental trading approach. This can prevent those huge drawdowns that can accompany blind (fixed) system failures. For those engineers who are reading this, consider a mechanical feedback system that dampens and focuses responses as a function of the speed of the transducers. If the feedback system is quick enough, it keeps up with the changes. But if it gets behind, it can go 180 degrees out of phase and tear itself up! It works the same with trading.

> 6. Trading a judgmental system is difficult, requires tremendous concentration, diligence, and self discipline.

> 7. There is tremendous challenge, fulfillment, and discovery in the entire trading process.

The bottom line is that there are certain inherent advantages and disadvantages to either market trading approach. I have chosen the judgmental approach. It is largely a question of matching your talents, your psyche, your financial resources, and your objectives with the challenges and advantages cited above. There is no way I've ever seen of avoiding a lot of work and a lot of stress. Prepare yourself for it, or if it's too hot, get out of the kitchen now!

SUMMARY:

ASPECTS OF JUDGMENTAL TRADING TECHNIQUES:

1. You can benefit from an extremely flexible market approach.

2. You will have a highly flexible personal schedule.

3. You have the potential for dramatic gains (or losses) quickly.

4. You will have potentials for extremely favorable win/loss ratios.

5. There is an absolute necessity for strict personal management.

6. There is an absolute necessity for a separate and adequate trading environment.

7. Relatively "small" amounts of capital can be adequate to achieve your goals.

8. Your focus on relatively few markets is not only acceptable, but preferable.

ASPECTS OF NON-JUDGMENTAL TRADING TECHNIQUES:

1. Poor win/loss ratios are the rule not the exception.

2. Historical hypothetical testing techniques are typically badly flawed for a wide variety of reasons.

3. Most non-judgmental systems ultimately fail; the aim is to attempt to use one while it's working.

4. Volatility breakout systems seem to best stand the test of time.

5. Multiple systems, traded over a wide variety of markets are necessary to smooth out the equity curve.

6. Relatively large amounts of capital are necessary for system and market diversification, as well as for the inevitable drawdowns.

7. If system and market diversification is achieved, large amounts of capital are employable.

8. Constant implementation of trading signals is essential, i.e. no breaks.

9. Locating adequate help to implement the system signals is a challenge in itself.

ASPECTS OF NON-JUDGMENTAL AND JUDGMENTAL TRADING:

1. An excellent life style is attainable if trading goals are met.

2. Fulfillment, challenge, and discovery are possible outcomes of the trading experience.

3. Stress levels can lead to utter destruction of your psyche and physical self, if not properly managed.

4. Financial ruin will accompany a frivolous or ill-advised approach.

5. The work load is awesome and unending; it must be properly managed.

6. You will have the potential to meet, and have as friends and colleagues, some of the best and brightest individuals on the planet.

POSITION TRADING VS. INTRADAY TRADING

You not only have to decide whether to trade judgmentally or non-judgmentally, you also have to consider the time frame that best suits you. Then, you need to be sure the time period you have chosen is best for the *approach* you are using.

With respect to applying the methodology this book teaches, it is easy. Essentially you apply the same general criteria to a five minute chart that you would apply to a monthly. Where *you* best belong is the tougher question.

My experience indicates that it's suicidal for a new trader, operating off the floor, to trade intraday. What's "new?" "New" is anyone trading *actively* for less than one year. If you are a part time or casual trader, you had better give it three to five years before going to intraday trading. The better question is however: What's intraday? My definition would be a trader who is actively observing price action during the day and making decisions based on what he believes is unfolding at that time.

A daily-based (or above) trader may choose an entry or exit point to be *acted upon during* the *next* day without being construed as being an intraday player.

What's a position player? The true answer is, it depends on perspective. To a floor trader, the five minute trader is a position player. To a daily-based trader, a weekly-based trader is a position trader, and so on. For our purposes however, we'll consider a position trader as daily-based or above. As you drop the time frame, the decision making time is compressed and the stress is increased. As you drop the time frame, the number of decisions increase radically. You have seven times the number of decisions going from daily- to hourly-based, 12 times the number of decisions, going from hourly to five minute. The opportunity is certainly accelerated, but I wouldn't expect a savvy boxing promoter to put a promising newcomer into the ring with Mike Tyson, just to see if he could learn to handle the big time more quickly. After all, he might lose more than his ears in the process!

DISADVANTAGES OF INTRADAY TRADING

1. You need experience - lots of it - with particular emphasis on order entry techniques and a thorough understanding of floor operations.

2. You need excellent brokerage and clearing services.

3. You have high overhead costs in software, quote delivery fees, equipment, and transaction costs.

4. So much of your time is taken up by the trading activity, you can't make money doing anything else.

5. Stress levels increase dramatically.

ADVANTAGES OF INTRADAY TRADING

1. You can trade many more contracts with a given amount of capital.

2. You will have many more trading opportunities than a position trader.

3. If your trading capital is severely limited and you are otherwise qualified, you have trading opportunities that will allow for much closer stops. Obviously the typical range on a five minute bar is smaller than the typical range of a daily bar. This point is really a variation of #1.

You should be able to restate the above to define the advantages and disadvantages of being a position trader.

AN ALTERNATIVE:

Given today's technology, there is a way to gain substantial benefits from a mix of the traditional paths. Here's how it works.

Get delayed intraday quotes from a quality vendor and have the capacity to display the bars graphically on an intraday basis. You can use, let's say 30 & 60 minute time frames, to help make daily-based decisions that can be acted on *during the next day*. The idea is to come home after work and make your decisions in the relative calm of the evening with the added accuracy and flexibility of intraday charting capability. You can set stops, entries, and such to be acted on during the *next* day. It may even be possible to set up some contingency orders, depending on your work environment and/or brokerage relationships. The advantages are substantial. You avoid the need for excellent brokerage services and a thorough understanding of floor operations. You avoid expensive software, on-line feeds, and equipment. You can make money doing something else. You can trade somewhat more contracts and have far greater opportunity than a traditional position player, and have closer stops. The analysis of your trading opportunities will be thorough. What is most important, however, is that you will operate with less stress than an intraday player, giving you an opportunity to grow into the trading experience rather than to be intimidated by its nature and seeming unpredictability. These are the options; pick your poison.

CHAPTER 2

PREREQUISITES,
GROUND RULES, & DEFINITIONS

□□

PREREQUISITES:

You should understand that this is not a book on basic technical analysis. It is a book about a modular and comprehensive trading approach that I have found to be both prudent and highly effective. Although I always attempt to start a discussion at a base level of assumed understanding, it is presumed that you have a working knowledge of certain well-known technical tools. If you do not understand Moving Averages, Stochastics, MACD, RSI, and the general appearance of price vs. time charts, there are a number of books listed in the reference section which you should consult before proceeding any further.

GROUND RULES AND DEFINITIONS:

Before we can hope to arrive at an understanding of my trading methodology, we have to be sure we're on the same page, i.e. that certain terms and concepts mean the same thing to both of us. To emphasize the *specific meaning* I attach to each of the defined terms, they will be capitalized throughout the text, i.e. Trend, Direction, Movement, and so on.

TREND:

A favorite question I ask when teaching is, "What is the current Trend of the S&P, bonds, or whatever?" Invariably I get the response, "Up, down, or sideways." Seldom am I asked, "In what Time Frame?" The question, "What's the current Trend of...?", without specifying the Time Frame is meaningless.

Below is a series of four charts. Let's use the Displaced Moving Average (DMA) overlaid on these charts as a Trend delineator. The length and type of this DMA is not the relevant issue. Our *definition of Trend* is the relevant issue. If we define an up Trend as a close above this DMA, and a down Trend as a close below this DMA, you can see the variety of Trend statements we can get on the close of the *same* day, February 28th. The bonds are in an up Trend on the 15 minute chart, but in a down Trend on the daily. The Trend is down weekly-based, but up monthly-based. If your trading methodology uses Trend as a part of its defining characteristic and you don't know *your* Time Frame, you are lost.

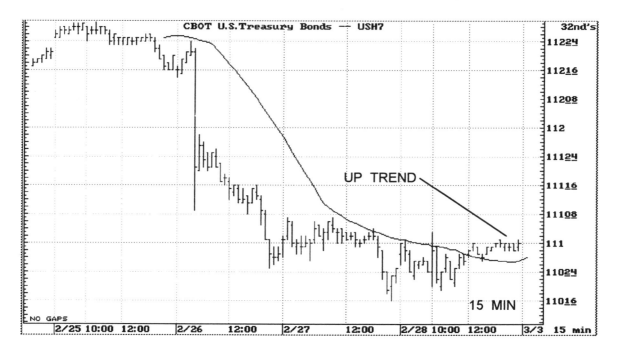

CHART 2-1

CHART 2-2

CHART 2-3

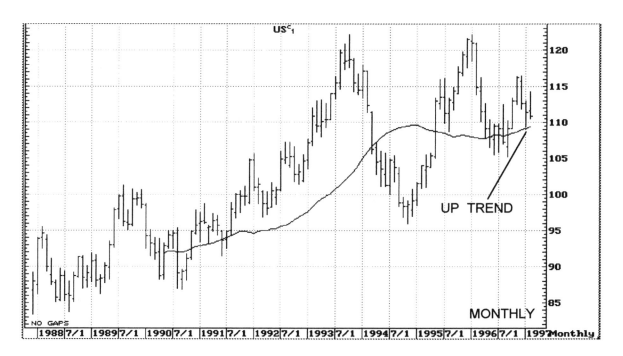

CHART 2-4

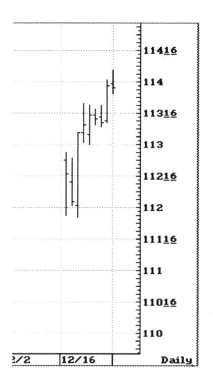

CHART 2-5

Furthermore, if I predefine the Trend by a set of indicators or by other criteria, I cannot make the determination of what the Trend is without the required criteria, regardless of how a chart may appear subjectively. Consider the following two charts. They are both daily segments of the monthly chart shown above. Chart 2-5 is shown without our predefined Trend indicator, the DMA. It would be easy to say the daily Trend is up, if you looked only at the sample of data shown in Chart 2-5.

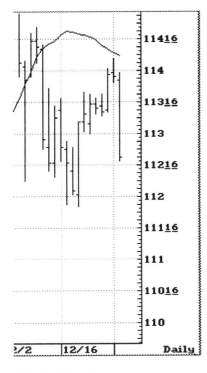

According to our definition of Trend, however, (Chart 2-6), this is merely a reaction in an ongoing *down* Trend. Perhaps you feel that given a longer data sample, you could have subjectively determined the Trend accurately in this case. Perhaps you could have. What about the next chart? What about the self-doubt you may encounter in the heat of action, and what about your ability to pull the trigger?

I use two Trend indicators and only two. They are Displaced Moving Averages and the MACD/Stochastic combination. Without them, I will not attempt to comment on the Trend.

My objective is to structure your thinking. One key to making a good *judgmental* trading method is to make as much of it as possible, *non-judgmental*.

CHART 2-6

When the market is going sideways, it is commonly said to be in congestion or without a Trend. The way I define Trend leaves little room for the term "congestion." It takes no leap of consciousness to see that congestion on a daily chart could be dramatically trending in another Time Frame, e.g. the intraday world. In the daily Chart 2-7 we have closing prices meandering above and below our Trend indicator. On the intraday Chart 2-8, using an increased vertical range, and the *same inputs* for the DMA, we have a solid Trend apparent. Unless there is some significant Movement in the market, I'm not interested in trading. Boring, meandering markets don't interest me. If shortening the Time Frame doesn't present a Trend, I simply stay away. Other methods that will help you to define "congestion," if you have trouble grasping the concept, will be covered in subsequent chapters.

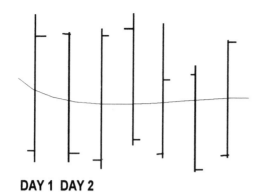

DAY 1 DAY 2

DISPLACED MOVING AVERAGE ON A DAILY CHART

CHART 2-7

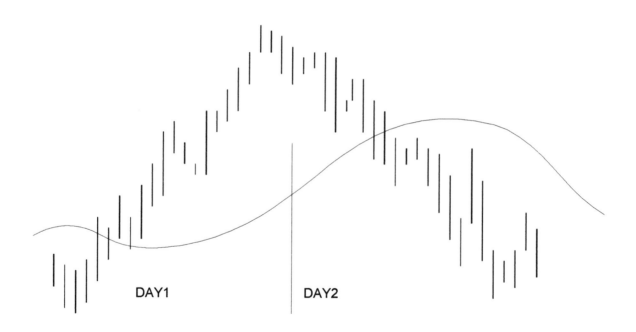

DISPLACED MOVING AVERAGE ON AN INTRADAY CHART

CHART 2-8

DIRECTION:

Direction is a concept, which like Trend, defines the Movement of the market, up or down. It differs from Trend in two important and distinctive ways. Firstly, Direction overrules Trend. That means if the Direction is up and the Trend is down, then the subsequent move of the market is expected and assumed to be up. We would interact with the market on that basis. Secondly, the criteria that determine Direction differ from those which determine Trend. Let me restate this point since the use of the term can seem confusing. Before reading this chapter, you had a concept of what the word Direction meant. You could look at a chart and say the Direction is this or the Direction is that. Forget your preconceived understanding of the word Direction. It will not serve you in the context of this discussion.

When I speak of the Direction of the market, it will be specifically defined and the ensuing price action should be highly predictable.

MOVEMENT:

The Movement of the market is a term encompassing Direction and/or Trend. One could say the Movement is up because this or that Trend is up, or the Movement is expected to be down because of this or that Directional Indicator. If the criteria is insufficient to make a definitive statement about Trend or Direction, you cannot make a definitive statement about Movement. Indicators or patterns do not *directly* define Movement. Only Trend or Direction define Movement.

FAILURE:

Another term you may have to redefine, is my use of the word "Failure." If the market experiences a Failure, it too will be specifically defined and subsequent market action should be highly predictable. Your previous understanding or your definition of the word Failure will not apply in this context. Failures are forms of Directional Indicators.

It is essential that you redefine some common terms, for in ten years of teaching traders, I have not found a better way to express these concepts. The concepts need explicit expression for thorough understanding.

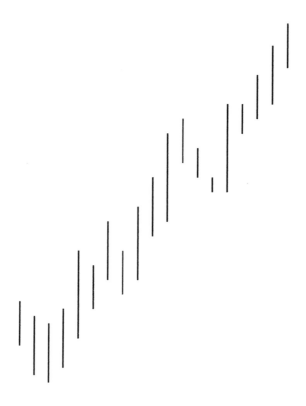

CHART 2-9

What is the Trend?

Hyper Hank: *Up, it's obvious, it's up! Let's buy it!*

Diligent Dan: *I don't know. You have defined how we determine Trend, but none of the indicators you use are on the chart.*

What's the Direction?

Hyper Hank: *You're nit-picking.*

Diligent Dan: *Same answer, don't know.*

What's the Movement?

Hyper Hank: *Come on guys, let's buy it before we miss out on the move!*

Diligent Dan: *Well, since Movement is dependent upon Trend or Direction, and since you haven't given us the information to determine either of these on the chart, I guess I can't say; but it sure looks like it's up.*

LEADING INDICATORS:

A Leading Indicator is one that indicates, before the market gets there, where support or resistance is likely to manifest. These types of indicators are typically not taught. They are certainly not used properly. Of those Leading Indicators that are available, there are few that are of significant value. The two Leading Indicators I find to be excellent are Fibonacci Retracement and Expansion Analysis, and a derivative of a Detrended Oscillator I pioneered in the early 80s, the Oscillator Predictor™. An example of a Leading Indicator that I do not find particularly useful, is time cycles derived directly from market action. Others however, Walt Bressert, and Peter Eliades among them, have done some excellent work on this subject. Some Leading Indicators that I find marginally useful are astronomical (not astrological) dates and certain time forecasts derived from Fibonacci counts. Those indicators, of course, do not predict support at a certain price, but rather that support will manifest wherever the price happens to be, at a given time.

LAGGING INDICATORS:

A Lagging Indicator is one that requires market action before the indicator turns. It confirms support or resistance rather than predicts it. In short, it lags market action.

Examples of Lagging Indicators are Displaced Moving Averages, Standard (non-displaced) Moving Averages, Stochastics, RSI, Trend lines, etc. At the risk of confusing the issue, it could be argued that certain of the above-mentioned indicators are coincident indicators, i.e. they indicate not before, not after, but simultaneous to market action. Understand that generally Leading Indicators are thought to be better than Lagging or coincident indicators, since they give you warning or precognition that certain steps should be taken. This is in contrast to confirmation that you should have done something yesterday or 15 minutes ago. Proponents, or teachers, of certain of these Lagging Indicators may see advantage in referring to them as Leading or coincident indicators to achieve a higher, perceived status for them. The truth, as I see it, is that one should use the best of both Leading and Lagging Indicators, and combine them in a way to achieve greatest success.

Let's explore this idea a little more deeply. A reasonable case could be made that once a Trend line is formed, it, too, is a Leading Indicator, i.e. from that point on, any approach of price to that indicator could be viewed as potential support.

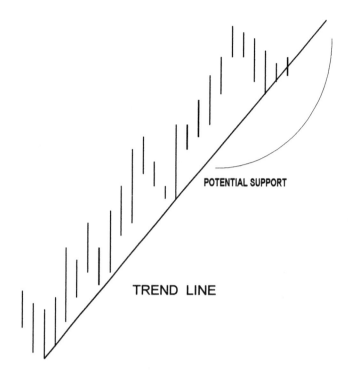

CHART 2-10

The same can be said for the Trend containment capability of a Displaced Moving Average.

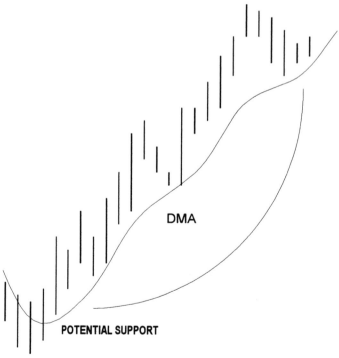

CHART 2-11

Okay, a case could be made. But why turn an excellent Lagging Indicator into a fair Leading Indicator as in the case of a DMA? Or why turn a fair Lagging Indicator (Trend line) into a poor Leading Indicator when we can use the best of both?

We could get into more contradictions and semantics on the subject of Leading, Lagging, and coincident indicators as an intellectual exercise, but that's not the purpose of this book. We will have enough of a basis for an intelligent approach to what will follow so I'll leave the topic here.

LOGICAL PROFIT OBJECTIVES:

A Logical Profit Objective is an area of price where one is likely to find significant resistance to further Movement, e.g. significant orders placed against one's existing position. It is not a place to necessarily reverse your position, and there's no guarantee the market will stop its Movement if it gets there. Logical Profit Objectives are *not* price projections. They are simply price points where the probability of continued price Movement decreases substantially. Fibonacci Expansion Analysis creates price points where one can take a logical profit if the market gets there. As long as the Trend or Direction (Movement) is intact, they will be fulfilled. In a strongly running market, it is likely all profit objectives will eventually be achieved, usually after tradable retracements. It is not advisable or helpful to see these points as projections *destined* to be fulfilled.

TIME FRAME:

While there are essentially an infinite number of Time Frame charts one can use to trade, I choose 5, 30, and 60 minute, daily, weekly, and monthly. Occasionally I'll look at 1 or 3 minute charts, but getting below a 5 minute chart is asking for trouble, since you begin to butt heads with floor traders at those levels. As discussed earlier, the shorter the "world" you live in, the better your brokerage and clearing services must be. The reason I don't trade 7 or 19 minute charts, or a 25 minute Time Frame to "beat out" the 30 minute players (an absurd concept) will be covered in CHAPTER 5 (The MACD/Stochastic combination).

CONFIRMED AND UNCONFIRMED:

If all the evidence is in on a given signal, it is said to be Confirmed. If we are waiting for some last piece of evidence the technical signal is Unconfirmed. An example of a Confirmed signal would be the *closing* of a price above a given Moving Average. An example of an Unconfirmed signal would be that same signal prior to the close.

MISTAKE:

A Mistake occurs only when you go against your Trading Plan, when you do something and you know better. A Mistake and a large loss may or may not occur simultaneously. In fact Mistakes can sometimes produce significant gains. If your Trading Plan is poorly defined, or you lack experience in the formulation of a Trading Plan, Mistakes will be hard to define. With experience, your Trading Plan will take form and Mistakes will be identifiable. It is critical to know when a large or small Mistake has been made, since the number and extent of Mistakes you make is *the best way* you can judge your progress - or lack of it.

TRADING WELL:

At certain times of the year we all do things better than at other times of the year. Whether it's your golf game or the way you interact with others, it's critical to be able to identify when things are flowing well. In golf, you can look at your score to get a measure of how well you are doing. In trading it's more difficult. Today's evaluation of *profit and loss* doesn't tell you much.

You are Trading Well when you are following your Trading Plan. A trader who does not Trade Well is, by definition, making mistakes. If he does not take corrective action according to his Trading Plan, he will never leave the game with profit on a *consistent* basis. You can always change your Trading Plan if you don't realize acceptable profit, but if you don't Trade Well, you are lost! Sell grapefruit, build houses, or remain retired. Don't trade!

TRADING PLAN:

A Trading Plan is a set of rules by which your trading is governed. They can be formulated with some flexibility as in a judgmental trading approach, or with absolute rigidity as in a non-judgmental trading approach. Even a judgmental approach should contain as many rigid rules as possible, so you will know with near certainty when you are Trading Well (not making mistakes) and when you are not.

SUMMARY:

Now we have the building blocks for our foundation. In CHAPTER 3, I'll put the foundation firmly in place. All of the concepts taught in the remainder of this book will fit within the confines of this structure.

CHAPTER **3**

THE ESSENTIAL COMPONENTS

OF A SUCCESSFUL TRADING APPROACH

MY TRADING PLAN INCORPORATES THE KNOWLEDGE AND IMPLEMENTATION OF :

1. *MONEY AND SELF-MANAGEMENT*

2. *MARKET MECHANICS*

3. *TREND AND DIRECTIONAL ANALYSIS (LAGGING AND COINCIDENT INDICATORS)*

4. *OVERBOUGHT/OVERSOLD EVALUATION*

5. *MARKET ENTRY TECHNIQUES (LEADING INDICATORS)*

6. *MARKET EXIT TECHNIQUES (LEADING INDICATORS)*

Let's take a closer look at each of these aspects individually and how they will be handled in this book.

1. MONEY AND SELF MANAGEMENT

I've already alluded to certain critical self-management techniques in CHAPTER 1. There's a lot more that you need to know[1]. In the reference section, I've made mention of some alternative sources of information that may be useful to you. Finding good material on market psychology will likely be easier than getting good, *practical* money management material.

Regarding successful personality profiles, I can add to the above reference material that successful traders are typically confident, self-made individuals, who can handle criticism and loss without defensiveness. There is no room in their minds for jealousy, envy, or emotional insecurity. It took me a long time to realize why nearly all my friends were traders and why I like being around traders so much. I don't know a better way to put it other than that they are *real* men and women. I'm not saying that traders must be perfect individuals. I am saying that the demands of this business substantially exceed those of other endeavors. If you suffer significantly from any of the personality flaws mentioned above, first iron out those flaws, then learn about trading specifics.

In any discussion of personal management and significant amounts of money, I guess it would be fair to comment on the people you'll run into as a part of your trading experience. We all know that there are thieves and unscrupulous individuals in any field. Such individuals exist in this business as well. I can honestly say, however, that in almost 30 years in and around this industry, I've run into only two. One has been indicted and the other is likely to be at some point. Those thieves that exist on the floor are straightened out by their (floor) colleagues. The motivation to clean things up is not moral, it's practical. There is only so much money to go around. If someone consistently gets too much of it unscrupulously, the others perceive this person as a pariah and find *very* effective ways to rid themselves of his presence.

---

[1] The FIBONACCI, MONEY MANAGEMENT, AND TREND ANALYSIS, *in home trading course* delves into the personality profile of successful traders. This portion of the taped discussion is approximately three hours long. It examines each aspect of successful thinking. More time is spent thoroughly covering money management techniques, an understanding of ruin (gambling) theory, as well as aberrant runs. The course explains how to handle margin during winning periods and losing periods. It discusses positive and negative expectation games and how your trading is defined by the nature of the methodology you are using. There is also a section that contains more basic information on how to open a futures trading account, how to choose a broker, and so on. Since these topics are well documented and thoroughly covered, they will not be repeated here. The appropriate audio portions referenced above, along with sections of the accompanying manual may be broken out of the full course. This way no one will feel forced into buying material that is perceived to be redundant.

2. MARKET MECHANICS:

I'm talking here about an understanding of the floor, of the actors on the floor, of order entry, floor rules, and practical issues...X'd trades for example[2].

If you were in the auto wholesale business and you didn't understand the insurance companies, the auctions, the dealers, the mechanics of the sealed bid, and such, how far would you get? If you were into real estate and you didn't understand the mechanics of financing and escrow procedures, how far would you get?

I run into traders all the time who know nothing about the floor, yet they use the floor six times a day trying to make money. Wake up! You need a reality check! Know the mechanics behind the business at which you are attempting to make a living.

In private seminars, I stress the importance of pit knowledge. I tell my students to get on the floor of the S&P for at least six hours. I want their legs to ache and I want them to see pit traders get spit on, trampled, and possibly even get into fights. I want them to feel the disappointment of the local who can't get an order off to close his profitable three-tick trade. I want my students to see the frustration of "out trades" in the making. Trading is a tough, risk management business. How can you manage a risk you are unaware of? How can you dispel misconceptions without experience? Floor traders are your colleagues. You do yourself a disservice when you make them your enemy.

What if you're not an intraday trader? That's fine. You don't need to know anywhere near as much about the pit as someone who trades intraday. You do, however, need to know more than 90% of the people out there who think they are going to pay their bills out of their trading profits. You need to dispel misconceptions that act as a crutch for impractical behavior.

This is a hot topic for me. When I designed my FIBONACCI, MONEY MANAGEMENT, AND TREND ANALYSIS in home trading course in 1988, I thought I had given the prospective trader everything he needed to get the job done. By relating a variety of personal experiences, I addressed this issue, but only so much could be done in a two-day workshop. I realized more needed to be done and I will endeavor to fill this void at some point in the future. I have identified the people I wish to have participate in such a work, but so far I have not been able to convince them to help me on this project. I will make constructive and appropriate reference to aspects of floor mechanics as we progress in this book. However I strongly recommend that you make it your business to seek out and befriend experienced traders, both on and off the floor. Once you have gained their trust,

---

[2] Joe DiNapoli , "THE X'D TRADE (or Where's My Fill?)," *Technical Analysis of Stocks & Commodities* magazine, March 1995, page 88.

they will freely discuss the anatomy of the floor and you will be rewarded for your efforts throughout your trading career.

I hope you are not unduly disappointed with references to other sources of information regarding items one and two above. Be sure you check the bibliography and reference sections. The rest of the trading approach is fully and explicitly covered in the following pages.

3. TREND AND DIRECTIONAL ANALYSIS
CHAPTERS 4, 5, & 6

4. OVERBOUGHT AND OVERSOLD EVALUATION
CHAPTER 7

5. MARKET ENTRY TECHNIQUES (LEADING INDICATORS)
CHAPTERS 8, 9, 10, 11 & 13

6. MARKET EXIT TECHNIQUES (LEADING INDICATORS)
CHAPTERS 7, 8, 9, 10, & 11

IMPORTANT POINTS TO NOTE:

While it is not a formalized part of my Trading Plan, I strongly recommend you complete trade evaluation summaries. This consists of a written evaluation or self-evaluation "diary", which enables you to gain perspective and reinforce the concept of Trading Well.

The remaining chapters of the book will deal with implementation of my Trading Plan, using a variety of specific examples.

SECTION 2

CONTEXT

When I get into an aircraft, I like to see some gray hair on the Captain. It gives me confidence and puts a smile on my face as I settle down to my first drink. My assumption is that he's been around a while and likely to have seen some spooky times. It's no different with trading. The egotistical traders among us might hide the Grecian Formula®, but the experience is still there. We juggle dynamite when we trade futures. While a trader could certainly use the concepts taught in this book to enhance his magnesium-like glow, the purpose of this book is rather to allow a trader to profitably survive. Longevity in this arena tells you who's made the grade.

By scrupulously identifying the CONTEXT of your trade, you know *up front* your risk/reward criteria. You know *up front* whether or not you wish to participate in a given trade or wait for the next opportunity which is always just around the corner. Remember...

Loss of opportunity is preferable to loss of capital!

CHAPTER 4

TREND ANALYSIS
DISPLACED MOVING AVERAGES

□□

GENERAL DISCUSSION:

There are many types and varieties of Trend delineation techniques. Trend lines and Moving Averages are among the most commonly used. There are also Displaced Moving Averages, Moving Average Bands, Deviation Bands, MACD, RSI, Stochastics and so on. There are almost as many ways of defining "Trend" as there are traders. One of my better students (Ed Moore) uses Fibonacci techniques to define "Trend." In inexperienced hands this could be a way to turn a great Leading Indicator into a poor Lagging Indicator. In Ed Moore's case, there's enough experience to draw from to make this transition useful. Another old time pro I know simply looks at price either above or below the open, to establish an up or down Trend respectively. While I like the simplicity of this technique, I think there's a real lack of quality in this method.

I use two specific methods for Trend delineation and only two. They are:

      1. Displaced Moving Averages
      2. The MACD/Stochastic combination.

In this chapter, we will limit the discussion to Displaced Moving Averages. The MACD/Stochastic combination will be covered in CHAPTER 5. The more exotic techniques for defining market Movement (up or down) are contained in CHAPTER 6, Directional Indicators.

If you haven't fully understood the term "Trend" as defined in CHAPTER 2, please restudy that section. If you are interested in how I came to use Displaced Moving Averages, you may refer to CHAPTER 1.

DISPLACED MOVING AVERAGES

Displacing a Moving Average forward in time offers several significant advantages to the trader.

1. It lets you know what the Trend delineation point or price number will be "N" number of periods ahead of time. Knowing where this point is, ahead of time, helps you to plan your market strategy.

2. By using the "proper" number of periods for calculation of the Moving Average and the "proper" displacement amount, DMAs tend to reduce whipsaws and "cup" or contain market action in ways that are very helpful to traders.

3. Certain DMAs are extremely useful in defining patterns, as shown in CHAPTER 6, Directional Indicators.

After many years of research spent selecting the proper length and displacement amount, I have arrived at three DMAs. They are:

- The 3 period simple Moving Average of the close, displaced forward three periods.

- The 7 period simple Moving Average of the close, displaced forward five periods.

- The 25 period simple Moving Average of the close, displaced forward five periods.

For brevity, they will be shown as follows:

                                3X3
                                7X5
                                25X5

The periods I use are Daily, Weekly, and Monthly. Quarterly and Yearly periods work equally well, but I seldom look at those Time Frames.

I've been teaching the use of DMAs for over 11 years. I've answered hundreds of questions on the subject. Since the same questions come up time and again, I think it would be useful to review them.

FREQUENTLY ASKED QUESTIONS:

What do you mean by "displaced forward in time" and how does this help reduce whipsaws?

Rather than plotting a given Moving Average, calculated today, on today's date, you simply plot the identical value at a different, later date, hence the term "displaced." The displacement is on the time axis, not the price axis. For the visual learners among you, the arrow in the following chart shows that the same Moving Average is simply placed forward in time. All calculations remain identical. For the mathematical types, Appendix A contains a table showing the calculations and where the respective values are placed.

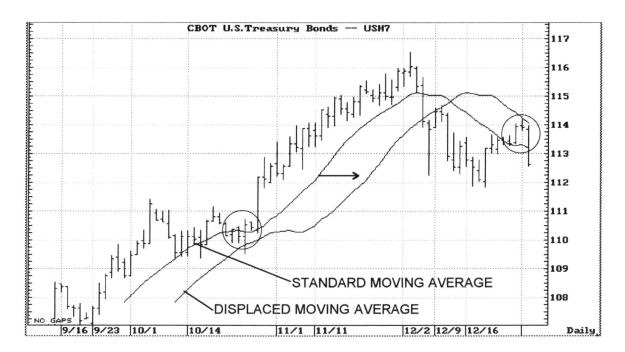

CHART 4-1

The next Chart 4-2A, shows a mathematically weighted Moving Average, (i.e. various "weights" given to different periods) plotted with no displacement. In other words, the chart shows a standard weighted Moving Average. The number of periods or the character of the weighting is not significant to the discussion.

Chart 4-2B shows the identical Moving Average displaced forward five days. The displacement amount could have been two, three, 10, or minus 10 days, weeks, or months. It is as if you took a piece of tracing paper that had only the Moving Average on it (no price bars), and slid the Moving Average left or right horizontally the desired amount (displacement).

Now, let's assume a basic Moving Average crossover, non-judgmental, always in the market system, where one would buy on close above the MA and sell below it on close. It's easy to see how many times one would have been whipsawed by the non-displaced MA as opposed to the Displaced MA.

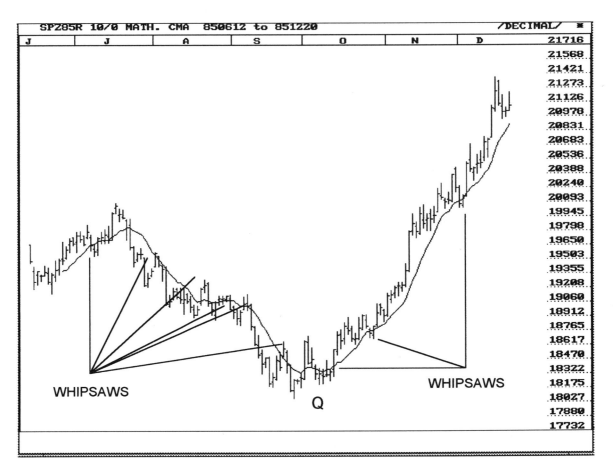

CHART 4-2A

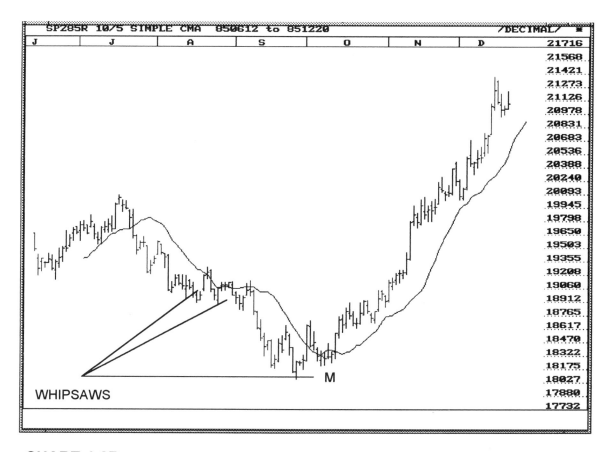

CHART 4-2B

For those of you who are attempting to build non-judgmental systems using Standard Moving Averages, try this additional variable and see if you get better results.

ADVANCED COMMENTS:

As with most observations in this business, there's the obvious and the insidious advantage to note. Obviously if you are whipsawed less often, you'll have a higher equity. Not so obvious is the real life factor of *continuing to play the game*. By the time most traders reached Q (for quit) as shown above in Chart 4-2A, they would have thrown in the towel and gone back to system development. Of course, this occurs just before the most profitable run on the chart. Point Q on Chart 4-2B has been relabeled to M for money because that is what will be made. The real life truth is that a trader trading off of Chart 4-2B is much more likely to be there for the big gains that are to follow!

Even being whipsawed infrequently, as in Chart 4-3A, can leave an average trader with an emotional excuse not to reenter. Look at the profit you would have missed if the whipsaw in March gave you an excuse to stay out of the market.

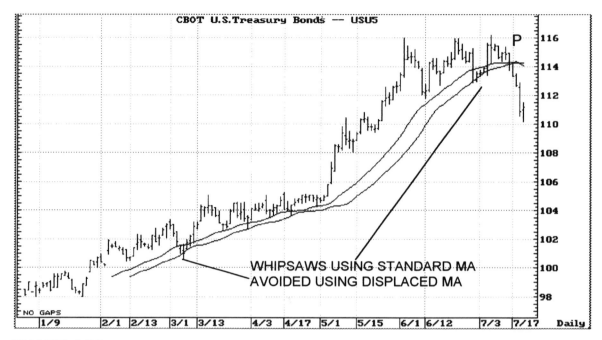

CHART 4-3A

Wouldn't using a longer Moving Average period accomplish the same function of preventing whipsaws and keeping you in the market?

Not really. Where a longer MA will likely provide fewer whipsaws, other properties change as well. See Chart 4-3A. Note how the displaced MA and the non-displaced MA approach one another when the market finally breaks. You maintain the same profit level, point P.

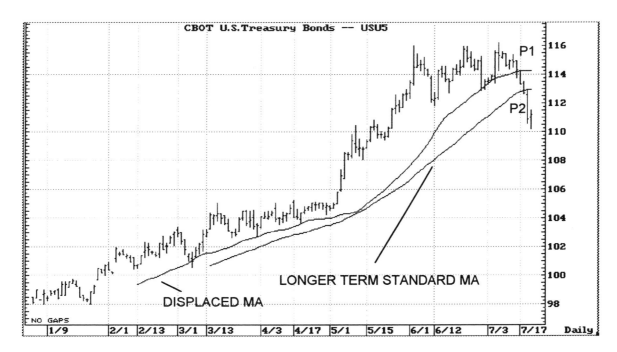

CHART 4-3B

The profit differential on Chart 4-3B is a different matter. Here we have a longer term standard MA. It falls dramatically away from market action. Therefore, a non-judgmental system whose signals are based on crossing a MA on close to take an open profit (P2), is likely to give back significantly more than the shorter period DMA would give back (P1).

How would you trade this?

I wouldn't. Trading any given market will come only after this book has thoroughly covered all the aspects needed to trade. The examples contained herein are for your understanding of the segment covered.

What did you mean about knowing the price delineation point 'N' periods forward in time?

The term 'N' refers to the displacement amount. If we're talking about daily periods, the 3X3 is displaced three days forward in time. You know the DMA value of the current day, two days, and three days ahead of time, i.e. the price point that will delineate trend. If there was no displacement (N=0) then you wouldn't even know until the *close* what the Moving Average value was for the *current day,* since you need that value to calculate the Moving Average.

Years ago, traders used the opening price rather than the close to calculate the Moving Average value, so they would know ahead of the close what the MA was for the *current* day. I believe the first series of workshops I gave in 1986 and 1987 discussing the advantage of DMAs was instrumental in the demise of this practice.

What about exponential MAs, weighted MAs, or "back-deviated Maxwell convoluted" MAs? Would they work better? Do the Displaced Moving Averages you use work in all markets?

You're welcome to try them. I arrived at the DMAs cited above over a period of about two and a half years, cranking them out on a CPM based computer using an 8088 chip. I studied thousands of charts, in all manner of markets, in all types of conditions. I tried every kind of MA I could imagine and have programmed. In those days there was no commercially available software I was aware of that created DMAs. To accomplish this task I needed to create a graphics package that would displace a Moving Average. The result was the first incarnation of the CIS TRADING PACKAGE, programmed by George Damusis. My research revealed no advantage in more complicated DMAs, over simple DMAs. Therefore in keeping with my primary directive, i.e. keep everything as simple as possible, I stuck with simple DMAs.

It's also useful to understand that I did *not* go through this optimization process, so popular with many of the computer junkies and cerebral types. Instead, I exhaustively examined market after market, to see what I, as an experienced trader, could live with emotionally and reasonably expect to be profitable. Could I do a better job today? I doubt it. More number manipulation power is not necessarily better. Besides, I doubt I'd have the fortitude to take on such a job today. Even if I were to make this *single aspect* of my Trend analysis technique 5% better, would it significantly influence the bottom line? I think not. As you will see, Trend analysis is filtered and acted upon by subsequent powerful techniques. Remember the old axiom, if it ain't broke, don't fix it. Don't misunderstand me on this point. Research is terrific. You can learn a lot from it, and you should try to improve on my work if you are so inclined. I would suggest to you, however, that you test your results across all the markets. The DMA values and calculation methods used should have *universal applicability* (overseas markets) as well as the ability to *stand the test of time*.

Below is a highly compressed chart of the daily UK long bond (gilt) with both the 25X0 and 25X5 shown. This chart illustrates price action in both trending and consolidating periods. Arrows have been placed at points where the 25X5 contained Trend but the 25X0 did not.

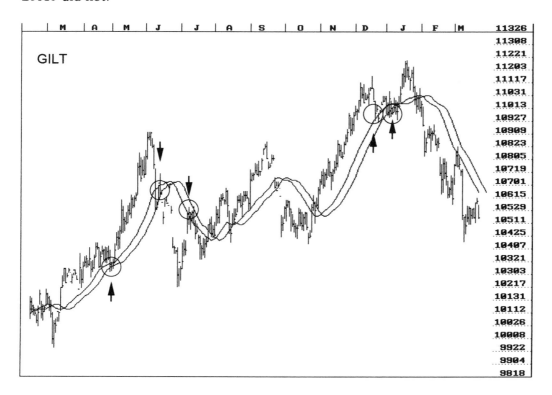

CHART 4-4

You do not use Displaced Moving Averages on intraday charts. Why not? Don't they work?

They work very well, but I have a technique that (for me) works better. It's the *MACD/Stochastic combination*. Many of my clients use DMAs on intraday charts and rave about their effectiveness. You, of course, are free to try them. They are simpler to employ than the MACD/Stochastic combination.

The following chart shows how well an intraday 3X3 contains thrusting moves on the 30 minute S&P.

CHART 4-5

So how do we define the "Trend"?

Very simply. If the close is above the Displaced Moving Average you choose, the Trend is up. If the close is below it, the Trend is down. If you use DMAs to non-judgmentally enter or exit a market, I'd suggest giving it a tick or two, through the DMA on close, before acting, particularly on the longer term 25X5.

What if the price is above it now, at mid-day, but yesterday's close was below it?

In that case, the Confirmed Trend is down and the Unconfirmed Trend is up.

Why do you use three value sets for the DMA?

The 3X3 is for the short term and is extremely useful in thrusting markets.
The 7X5 is a longer-based DMA that many have found useful in equity market analysis.
The 25X5 is my long term DMA.

What if the closing price is below the 3X3, but above the 25X5?

Then the short term Trend is Confirmed down and the long term Trend is Confirmed up.

Below is an example of both DMAs on the daily German bond (bund).

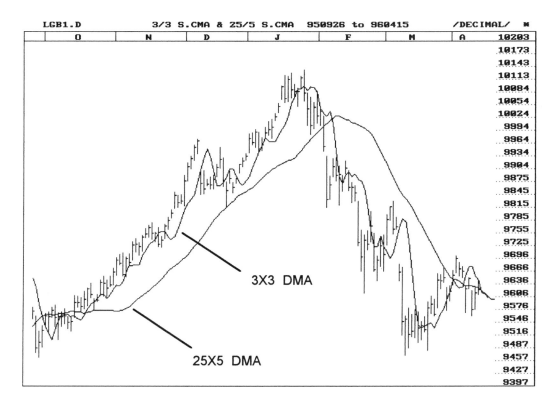

CHART 4-6

How you would react or how you would play this would depend upon what Time Frame player you were. If you were an hourly-based player, you would be very interested in the daily 3X3 and you'd be aware of the 25X5. If you were a weekly-based player, you would be interested in the 25X5 on the daily or perhaps the 3X3 on the weekly. One caveat here. If the weekly-based trader was aware of a Directional Indicator created from a daily-based 3X3, he would take notice. I'll explain this in detail in CHAPTER 6.

Is it okay to take a position on an "Unconfirmed" signal?

Absolutely, I do it all the time, but at the close of the period you had better have confirmation in the Direction of your assumptions or you say "adios" to the trade. If you do not, you have made a major Mistake. Also, on the subject of Mistakes, you *never,*

ever, change the reason for a trade. If you entered a trade on the basis of some criteria and that criteria is negated, you don't look around for other criteria to justify your position. If you do, you have made a serious Mistake. *Close the trade*, and if you have other criteria to be in the trade, re-open it, on that basis. Pay the commission. The long term psychological considerations far outweigh the costs involved. Besides, once you're out of a trade and take a fresh look at it, the reentry criteria may not seem nearly as compelling.

You referred to cupping market action. Is this really the same thing as minimizing whipsaws?

Yes, one of the most common techniques misused by traders is that of tightening up on their stops too quickly. The idea sounds good, but most traders don't have good stop placement techniques in the first place, much less any idea about *when* to tighten them.

Consider the following daily chart of the Canadian dollar.

CHART 4-7

From March through the beginning of May, the market is primarily in a strong up Trend as defined by the 3X3. We would therefore be playing the long side, by buying dips, likely on an hourly chart, and selling at specific objectives. These dips and Objective Points are defined by Fibonacci work which will be covered later.

When we reach the top, we have a sustained break through the 3X3 to the downside and the 3X3 contains or "cups" the ensuing rally back toward the top. Look at the expanded view of the top, Chart 4-8.

CHART 4-8

If we tightened our stops too quickly, after getting short on day A, we would be stopped out on day B or C. If we used the 3X3 as a guide to Trend delineation, we would have no reason to be concerned about our short position, unless we experienced a significant close back above the 3X3. In later chapters, we will cover stop placement techniques thoroughly. For now, what's important is that you understand that the 3X3 is giving the market an opportunity to breathe. The rally back toward the previous top provides a way for those who have provided market liquidity to the sellers, to unwind the long positions they have accumulated, and likely go short themselves.

As you study the concepts defined in this book, you will plainly see that markets are *stable and tradable* if they have reactions to any significant price Movement. They are *unstable and dangerous* to trade if there are sustained and violent moves in only one Direction. This is due to the fact that those professionals who have provided liquidity to the markets are the wrong way!

CHAPTER 5

TREND ANALYSIS
MACD/STOCHASTIC COMBINATION

ᴼᴼ

GENERAL DISCUSSION:

Like most traders new to the Stochastic, I was befuddled for a time by its movement, and dismayed by my initial attempts to use it. Fortunately the equipment I used at that time to display the Stochastic was the CQG TQ20/20™. I say fortunately, because the TQ used a *modified type* of Stochastic, rather than what was *said* to be the standard Lane Stochastic. Some time later I learned there were differences in Stochastic formulas. Had I attempted to apply this indicator using a different Stochastic formula than what was programmed in the TQ20/20™, my learning curve would have been much steeper. This is because the indicator would have been much more difficult to apply and interpret.

PROGRAMS, PROGRAMMERS, AND PROBLEMS:

At the risk of having your eyes glaze over from boredom or creating a brain embolism from strain, we're going to digress a bit and discuss some significant issues we, as traders, encounter while attempting to utilize trading tools. The Stochastic, and to a lesser extent the MACD, give us a perfect setting for this discussion. We'll consider the Stochastic first.

George Lane[1], the originator of the Stochastics, observed that the closing price within the range of a bar had significance relative to future price development. After considerable

---

[1] George Lane, *Investment Educators*

diligent effort, he derived a formula that quantified this belief. That seems straight forward and simple enough, but in the real world of trading software it's anything but straight forward and simple. There are disturbing variations of Stochastic formulas floating around and specified in reference material. Even talking with George himself who is one of the most knowledgeable, generous, and gentlemanly figures associated with the business, I have not found a simple means of getting from the *original formula* to what we as traders are faced with using now. So here's how we'll proceed. I will *try* not to bury you in complex math. There are a variety of equations in Appendix E for the mathematicians and programmers among you. I will also give you the option to skip ahead a few pages to the Preferred Stochastic, if your only interest is in knowing what I use. By doing so you will eliminate going through a somewhat tortured discussion. However, you will miss out on some of the issues we face utilizing trading software to make our trading decisions.

As an outgrowth of George's work came a variety of terms:

Lane Stochastic
Raw Stochastic
Fast Stochastic
Slow Stochastic
Modified Stochastic
The Stochastic

WILL THE RIGHT STOCHASTIC PLEASE STAND UP:

When traders purchase a graphics package from let's say, Trade-Em-Quick Software Inc., we can see that they have "The Stochastic" available as one of their *preprogrammed* indicators. Great, we're happy, since that's one of the indicators that we've read about and want to use. But, which Stochastic is "The Stochastic"? If we don't know enough to ask some very pertinent questions to some hopefully informed and responsible sales people, we haven't any idea of what we are *really* getting! So, let's try to learn something about Stochastics and *let's be worldly about software and how it is developed*.

LANE (RAW) STOCHASTIC:

As far as I'm concerned, all Stochastics can rightfully be called the Lane Stochastic since they all emanated from George Lane's research. Lane Stochastics, i.e. all of the Stochastics we will discuss, have two lines: a fast moving line %K, and a slow moving line %D. There seems to be some agreement between the different Stochastics formulas for %K of the Fast Stochastic, sometimes referred to as the Raw Stochastic, so we'll start with it and I'll include the equation here.

%K, Fast (Raw)Stochastics:

%K = 100 [(C-L_n)/(H_n-L_n)
where:
C is the latest close
L_n is the lowest low for the last n days
H_n is the highest high for the last n days

FIGURE 5-1

The calculation of %D, the slow line, is where a large number of the problems arise. The slow line %D, is a smoothed version of the fast line, but there are a variety of ways of smoothing a line. There's the number of periods we can use to smooth a line, as in a *five* period Moving Average, or a *ten* period Moving Average. Then, there's the *type* of Moving Average we can use. We could use a *simple*, or an *exponential* Moving Average for example. Since there are a variety of ways of smoothing a line, there are a variety of different Stochastics.

FAST STOCHASTICS:

If we use the formula in Figure 5-1 for %K and smooth it by using a three period *modified* Moving Average (MAV), we have the %D line of the Fast Stochastic. In George Lane's Stochastics article[2], he used an example from the TQ20/20™, which used *this type of smoothing* of %K to create the % D line. The TQ was also programmed to use the same type of smoothing to create the Slow Stochastic referred to in Figure 5-3.

%D (of the Fast Stochastic) = 3 period MAV of %K (of the Fast Stochastic)

FIGURE 5-2

Some software companies will use *other smoothing methods*, however, and still call the indicator the Fast Stochastic.

---

[2]George Lane, "Lane's Stochastics," *Technical Analysis of Stocks and Commodities,* May/June 1984.

SLOW (PREFERRED) STOCHASTICS:

Slow (Preferred) Stochastics are derived from Fast Stochastics.   If we take the %D computed above and rename it to %K, and then smooth this line by using a *three period modified Moving Average*, we have the new slow, Slow Stochastic line, %D. These two lines make up the indicator called the Slow Stochastic, created by a modified Moving Average smoothing. This is what I use ("Preferred").

> *%K (of the Slow Stochastic)= %D (of the Fast Stochastic))*
> *%D (of the Slow Stochastic)= 3 period MAV of %K (of the Slow Stochastic*

FIGURE 5-3

Some software companies will use *other smoothing methods*, however, and still call the indicator the Slow Stochastic. The formula for the modified Moving Average is shown below. The starting point (MAV_t) is calculated identically to that of a simple Moving Average.

MODIFIED MOVING AVERAGE (MAV):[3]

> $MAV_t = MAV_{t-1} + (P_t - MAV_{t-1})/n$
> *where:*
> MAV_t *is the current modified Moving Average value*
> MAV_{t-1} *is the previous modified Moving Average value*
> P_t *is the current price*
> *n is the number of periods*

FIGURE 5-4

If a *simple Moving Average* is used instead of a *modified Moving Average* to perform the smoothing, you get a Slow Stochastic which is significantly less useful. In fact, I find it useless.

---

[3]P.J. Kaufman, *The New Commodity Trading Systems and Methods*, (New York: John Wiley & Sons, 1987)

MODIFIED STOCHASTIC:

If we start at the original agreed upon formula for the fast %K (Figure 5-1) and smooth that line by any means whatsoever, we have %K of the modified Stochastic. If we then take that %K line and smooth it by any means whatsoever, and call the result %D, we have the slow line of the modified Stochastic. It is likely you will find other definitions of the Modified Stochastic in reference material or software users manuals.

THE STOCHASTIC:

Display Trade-Em-Quick Software or Aspen Graphics™ or CIS TRADING PACKAGE or TradeStation® on your screen and there you have it, an indicator that is a Stochastic. What it is or how useful it is, is anyone's guess. Without diligent research, I wouldn't hope to define this term as the study it suggests can have such vastly different appearance, applicability and usefulness, depending on how the equations are manipulated in the software of your choice!

THE PREFERRED STOCHASTIC:

This is a new term. It is meant to reflect what *I use and find of benefit*. The equations cited above to produce the Slow Stochastic and the modified Moving Average achieve what I want. Other equations and references relating to the Stochastic are in Appendix E so as not to confuse this issue further.

The last time I looked, my Preferred Stochastic was called the Slow Stochastic in CQG, Inc., Aspen™, and our own CIS TRADING PACKAGE. It was not available in TradeStation® as a preprogrammed indicator, but could be created by inputting the proper equations in so-called "Easy Language™"(Appendix D). MetaStock™ defaults do not produce the Preferred Stochastic. You can change the defaults in MetaStock™ without inputting equations, to create the Preferred Stochastic.

When you study the examples of Stochastics on charts in this book, using the Aspen Graphics™ software, you will see the name *Modified* Stochastic rather than *Slow* Stochastic, even though *Slow* Stochastic is my Preferred Stochastic. Why? When I set up the study initially, I didn't trust that the programmers had computed the Slow Stochastics correctly. I therefore went to the Modified Stochastic study and set up the study myself, to duplicate what I knew were correct inputs. I then compared these values with what I knew were correct: our own CIS TRADING PACKAGE. After I did that, I compared the specific Modified Stochastic I set up in Aspen with what Aspen called the *Slow*

Stochastic, and found out their programmers did get it right. When we discuss "The Stochastic" in the body of this book, it will be my Preferred Stochastic.

As trading software evolves, it is likely that the Modified Stochastic will supplant all other forms of Stochastics, since by definition the Modified Stochastic can be adjusted to simulate all others. In that case, to simulate our Preferred Stochastic, the user would input four variables:

1. *eight* periods for consideration (eight days, eight hours, etc.)
2. *three* periods of smoothing for the fast line
3. *three* periods of the smoothing for the slow line
4. *modified* for the type of Moving Average to accomplish the desired smoothing

As if this level of detail does not complicate things enough, let me tell you about two other aspects you need to watch out for when selecting software packages and trading with the indicators they produce.

MARKET-ALIGNED VS TIME-ALIGNED BARS:

It is much easier to program time-aligned bars but they are not as good as market-aligned bars for analysis. Let's look at bonds as an example. Even though the bond market starts trading at 8:20 AM and ends its actual half hour at 8:50 AM, time-aligned bars would begin this market at 8:00 AM, and end the first bar at 8:30 AM. In this example, the first 1/2 hour (8:00-8:30) *will have 10 minutes of actual data in it. The second half hour bar will have only 20 minutes of the first half hour's data in it, and 10 minutes of the second half hour's trading data in it.* Another example of time-aligned bars producing "erroneous" high, low, last data, would be an hourly S&P. In this case, an hourly S&P's first bar would contain data from 9:00 AM to 10:00 AM although data is not flowing until 9:30 AM! The second hour starts at 10:00 AM and goes until 11:00 AM, instead of correctly starting at 10:30 AM and going until 11:30 AM. With the high, low, last for these intraday charts being recorded "incorrectly," all studies calculated from them are obviously also incorrect. Don't let complacency put you to sleep on this one. Many traders have used studies generated from time-aligned bars for years, with substandard results. *Many of these traders are totally unaware of how these studies are being calculated. I suggest to you that the poor performance of indicators may be the result of the improper basis from which they are calculated, rather than the quality of the indicators, or the trader's understanding of their use!*

THE DATA SAMPLE:

The second, and again, not so obvious malady certain graphics packages suffer from, derives from the data sample they select to calculate studies. Let's say you reduce the horizontal (time) axis from 140 to 40 bars. If the studies you are using require in excess of 40 bars to produce accurate values, some vendor programming will be inadequate in that they will *calculate* only *from the sample shown on the screen*. Whether you look at 20, 40 or 400 bars on the screen, good graphics programs will give you the same values on the studies. These values *should not* depend on the number of days shown on the screen assuming, of course, that you have adequate data available on your hard drive to make the calculations accurate.

Loose talk about "this Stochastic" or "that Oscillator," without researching the formulas that are used to create them or the programming behind the creation of the bars from which they are calculated, can lead to the most disappointing results, *with no hint of where the problem lies!*

PROGRAMMERS AND UPGRADES:

On the subject of programmers, let's talk about what happens in the real world of software development. Let's assume you're the president of a software company and you're also a trader. You have a very stable piece of software you use daily for trading decisions but...it has a teensy, little bug in it. The number 8 in 1998 shows up a little too far to the right of the screen. You go to your programmers and say, "That's kind of annoying; can you fix it?" "Sure!" is the answer. You get the software back two months later and you find out they fixed the number 8, but they also "fixed" a "problem" one of them noticed in the Stochastic equation. Of course, *they didn't tell you about this "fix."*

I'd like to establish an industry-wide standard to ensure that when programmers do something and *don't tell you about it,* they lose a toenail. If you think that's harsh, consider a trade you've been planning for weeks, that should have netted you $20,000, but lost $10,000 instead. Why? Because the calculation of an indicator in your trading plan had been changed without your knowledge or approval. You tell me, *as a trader,* would you personally like to find the pliers, or would you just smile and say, "Please don't do that again?" If I had my way, there would be a lot of limping going on in trading rooms, at least initially!

The same goes for upgrades. The software company that has produced your trading software says they have this fantastic *new thing,* this *wonderful* study in their upgrade you couldn't give a gaseous burp for. You're *forced* to "upgrade," however, since they're no longer going to support that old software version you're working with . You find out *later* that in the "upgrade," they screwed up the continuous contract generator, created a bug in the cursor window, and your charts won't print correctly any more! When you tell them about this they say, "Don't worry, there's a new upgrade on the way - for *only* $195.00."

Besides undocumented changes, so-called upgrades can wreak havoc with your trading plan in other ways. Often, a default setting on files contained in the upgrade can overwrite settings *you may have painstakingly inputted.* I'll give you one example which may or may not apply to the software you use. Bid and ask prices are routinely reflected on your quote page. Most traders want this feature. Most traders, however, *do not want* bid and asked prices charted. If you select the "bid and asked not charted" option and this is overwritten by your upgrade, it may take months before you realize your charts are wrong! Meanwhile your indicators, D-Levels™, your highs, lows and lasts, will all be incorrect. You may even think you're due a fill by looking at your chart, when there was only a bid or offer beyond that price!

If you are new to this game take heed. I've been involved with trading software generation and use for 15 years and these issues are *real problems.* While I am awed by the talent of programmers, I am equally dismayed by some of their actions and the actions of the managers who direct their efforts. These individuals who may know as much about trading as squirrels do about Fourier analysis, can easily take it upon themselves to "improve" or inadvertently destroy our critical decision-making tools!

Notwithstanding the above, were it not for the incredible talent, persistent, dogged work of programmers, I would not have had an opportunity to develop the way I have as a trader. Without programmers, the Displaced Moving Average research, the Oscillator Predictor™ and the striking advantages of my FibNodes™ program would have remained as unfulfilled dreams. So, recognize that the advantages and costs of utilizing computers and human beings to program them, offer awesome benefits, as well as serious challenges. If you keep in mind that software engineers require the same level of *strict management and diligent oversight* as traders, the benefits should far outweigh the costs.

USING THE STOCHASTIC

In the early days, I started out with inputs of 14, 3, 3, but later settled on 8, 3, 3. Many of the initial problems I encountered with this study were solved when I internalized the concept of Trend being dependent on Time Frame. There is absolutely *no inconsistency* with the five minute Stochastic showing a "buy" while the half hour shows a "sell." The inconsistency, if any, is in the user's mind, by not knowing what Time Frame *he* is trading in, or lacking the experience to handle the speed of variations inherent in very short term, intraday trading.

In my early days trading futures, I used only the Stochastic to determine intraday Trend in the traditional manner. When the fast line crossed the slow line from below 25, and broke above 25, it signaled an up trend. When the fast line crossed the slow line from above 75, and broke below 75, it signaled a down trend. See Chart 5-1.

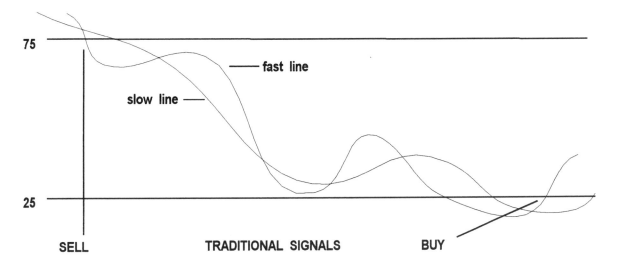

CHART 5-1

THE MACD (DEMA) STOCHASTIC COMBINATION:

During the mid 80s, Jake Bernstein and I conducted a seminar together. One of the topics he presented was his Dual Exponential Moving Average (DEMA)/Stochastic combination method. Over the years, Jake taught me many things but this particular technique, altered in a specific way, remains among the most powerful in my trading arsenal. The way Jake taught this method was to use the Stochastic in the traditional manner and *filter* it with the

DEMA buy or sell. In other words, the Stochastic and the DEMA had to both be on a buy, or both be on a sell, before a *confirmed* Trend signal up or down was given. So what was a buy on the DEMA? Better yet what is the DEMA? The DEMA is a derivative of Gerald Appel's MACD[4] (Moving Average Convergence Divergence) originally developed by Mr. Appel for analyzing stock trends. The MACD is, as Mr. Appel says, quite simple. You take the difference of two Moving Averages of price and create a Moving Average of that difference. The difference of the original two Moving Averages and the Moving Average of that difference, can be plotted as two lines, one fast and the other slow. The equations are in Appendix E.

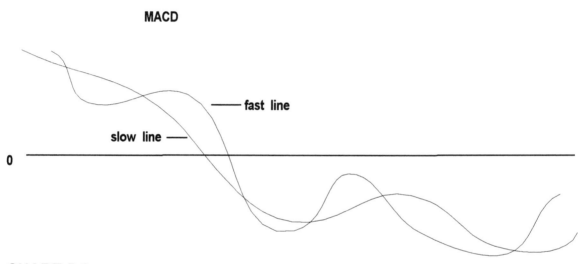

CHART 5-2

Notice, I've used the same two wavy lines on Chart 5-2 to show the MACD that I used to show the Stochastic in Chart 5-1. All I've done is change the scale, since the MACD (DEMA) oscillates about the zero line, while the Stochastic travels between zero and 100. In our work we will essentially *ignore the scale* for both indicators and simply observe the penetrations of the wavy lines.

Jake Bernstein took maximum advantage of the MACD by using specific *exponential* Moving Averages rather than whole number inputs as Gerald Appel did, hence the term DEMA [5].

---

[4] Gerald Appel, *The Moving Average Convergence-Divergence Trading Method* (New York: Signalert Corporation).

[5] Jacob Bernstein, *Short Term Trading in Futures* (Probus Publishing Company, 1987).

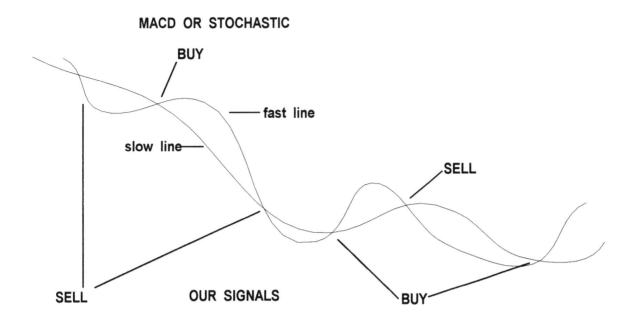

CHART 5-3

Like the Stochastic, when the fast line crosses the slow line from below, you get a buy. You get a sell signal when the fast line crosses the slow line from above. As much as I like to tweak things to achieve maximum advantage, I have never found any combination of inputs that matches, much less exceeds, those that Jake developed, i.e. .213, .108, .199. These exponential inputs can be simulated by "period" inputs of 8.3897, 17.5185, 9.0503, *if the software you are using is programmed to take "period" inputs and simulate exponential Moving Average smoothing.* Among those graphics programs I know of which accurately program this study are CQG, Inc., Aspen Graphics™, TradeStation®, and our own CIS TRADING PACKAGE. I'm sure others do as well, but I have not *confirmed* that fact.

If you're a "shoot from the hip" type and think I'm getting just a little too particular about these details, that's your prerogative. I must emphasize what I believe is important. It's your option to choose to ignore what you please. I'm not saying you will lose money if you don't follow this precisely. I am saying that you should know *what it is that you are doing and not make unwarranted assumptions.* Besides, we need lots of "shoot from the hip" types. These traders often provide us with the other side of our trade.

Okay, let's assume we have properly calculated and displayed the Stochastics and MACD studies (I will refer to the DEMA as the MACD from this point forward for clarity). Here's how I use them.

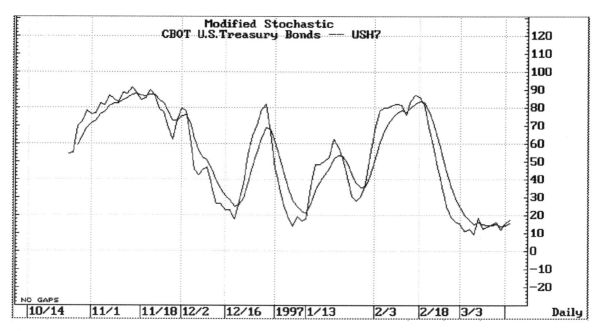

CHART 5-4

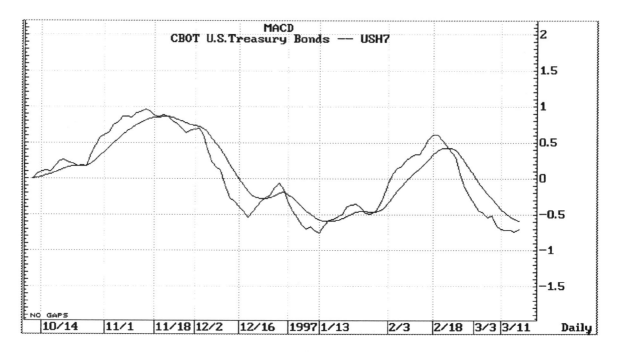

CHART 5-5

For my purposes, the MACD is the more reliable Trend indicator. I have continued to use Jake's numbers, thereby leaving it strong. I have deliberately weakened the Stochastic by inputting 8, 3, 3, rather than the stronger 14, 3, 3, originally used by Jake and many others. Above are Charts 5-4 and 5-5 of the Stochastic and MACD for the daily March

U.S. bond contract. Notice that the Stochastic study has a more ragged appearance than the MACD. The flowing lines of the MACD give us a smooth presentation of Trend. This is what we want! The number of buy and sell signals given by the MACD are infrequent compared to the Stochastic. By displaying these two indicators, one on top of the other, in separate windows, as in Chart 5-6, I can evaluate when to fade (go against) a market and when not to.

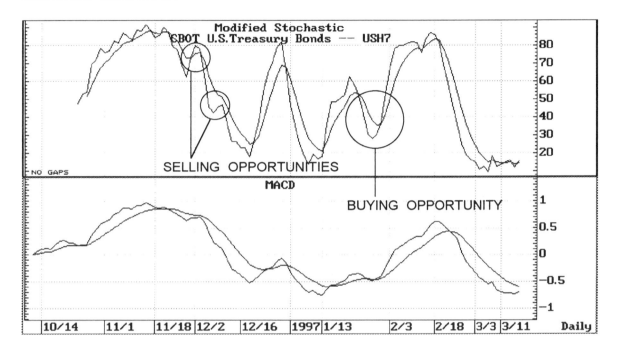

CHART 5-6

If both the Stochastic and the MACD are on a buy, the Trend is up. If the weak Stochastic signals a sell, I can buy the dip associated with that sell, as long as the (strong) MACD buy remains intact. Notice the solid up trend shown on the MACD between mid-January and mid-February. We had an excellent buying opportunity *against Fibonacci support*, as the weak Stochastic went into a sell, then got back in gear to the upside. The same can be said on the sell side for the period from mid-November through mid-December. What is more subtle but nonetheless opportune, is the MACD up trend to the far left of the chart. While casual observation suggests that the Stochastic remained in an up trend, real life trading was just the opposite! Why? During the period from mid-October through the November high, there were many times when the Stochastic signaled *unconfirmed sells during the day*. We can't see those sells by looking at this chart, because the close of the day is the point at which the (confirmed) indicator is computed and this is what we see on Chart 5-6 when we look back. Traders acting on those unconfirmed intraday signals in real time, however, gave us an opportunity to go long *when they sold*. We could then take profit at pre-calculated Fibonacci Logical Profit Objectives when the Stochastic went back into the buy mode and *their buy stops were hit!*

By observing the *combination signals* on the most frequently used charts, i.e. 5, 30, and 60 minute, daily, weekly, and monthly, we can get a bird's eye view of where the weak players are (Stochastic), as well as the position of the strong players (MACD). My aim is to buy dips (Stochastic sells) at Fibonacci retracement points in an up trend (MACD buy), or to sell rallies (Stochastic buys) at Fibonacci retracement points in a downtrend (MACD sell). In this way, I combine Leading (Fibonacci) and Lagging (MACD/Stochastic) Indicators, in such a way as to "safely" interact with price action. You should also note that the traditional requirement for a Stochastic signal to be at the extremes of 25 or 75 is ignored. Just like the MACD, *I require only a crossover of the fast line through the slow line for a signal.*

Chart 5-7 shows a five minute S&P in a thrusting down move. It's difficult to see the extent of the thrust, since the bars are confined to only one third of the chart. This was done so I could show you the action of the MACD and Stochastic. Initially, both indicators are signaling a sell. An interim price low is reached and the Stochastic goes into the buy mode. It is bringing in the weak longs and getting rid of the weak shorts. The down trend, as defined by the MACD, remains intact. Observing this type of action will show you how trailing stops set in improper areas give knowledgeable players a perfect opportunity to take away positions from weak players, i.e. to buy a dip or sell a rally in the direction of the prevailing Trend. Chart 5-7 shows how the buy stops are hit, driving the market up to Fibnode resistance. After the up move, the Stochastic gets back in line with the MACD and the market returns to its previous direction, perhaps to new lows. This type of action is repeated again and again in a variety of Time Frame charts. Just be sure you are involved in thrusting markets in order to help avoid possible whipsaws.

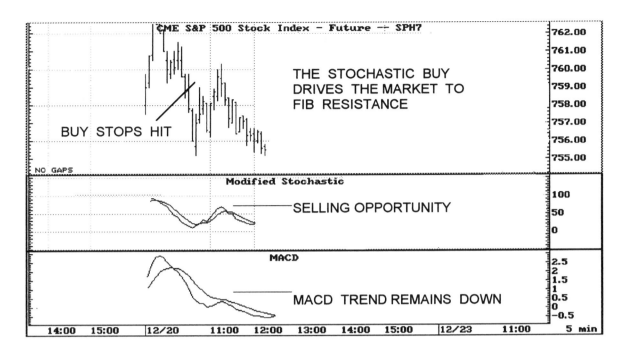

CHART 5-7

In teaching the use of this MACD/Stochastic combination signal, I typically break it up into levels of complexity and teach up to the level possible, depending upon the presentation setting and the experience level of the students. The above explanation involves level 1 (waiting for both indicators to agree before defining a Trend), and level 2 (the concept of fading the weak Stochastic indicator while positioning yourself in the prevailing Trend). Later we will have examples of level 1 & 2 utilization of this technique. Level 3 (anticipating or acting on an unconfirmed signal) will be discussed somewhat, while level 4 involves sliding your Time Frame and is too complex a subject to be adequately covered outside a classroom setting. To give you some idea, however, as in the example above, the half hour Trend would typically be on a sell *to further support our Stochastic fade on the five minute chart.* There will be more examples after we cover Fibonacci analysis.

Now, let's drop back, and look at this from a different perspective. If you think about this approach and consider the mathematics of the Stochastic, you will see how a market can be *made* to turn. Consider a large local or, more likely, a group of locals who are short the market. If they can hold prices at a given high for several bars (keep prices from going higher), it will force the (weak) Stochastic to turn south. The weak longs start selling their positions and weak shorts initiate new positions on the sell side. Now the locals (and we) can buy those sell orders. The locals can take their several ticks profit while we can position ourselves for the expected new high, or a move up to a Fibonacci expansion point. If we tried to buy stop the old highs instead of buying in on the dips, we would be at a point of high market slippage. Upon being filled we would have to suffer through yet another pullback, while the locals who have fed us the sell orders, endeavor to make a profit. If we buy the Stochastic sell and the MACD ends up breaking (giving a sell as well as the Stochastic), we know we're wrong and we take the next rally out. If we are operating on a short term Time Frame and have adequate experience employing this method, it is possible to break even, pick up a few ticks, or perhaps suffer only a few tick loss, *even when we are wrong!*

CHART 5-8

Let's take a look at a relatively simple example on daily crude oil, Chart 5-8.

We obviously have a thrusting, up trending market as defined by the 3X3. If you played this market primarily from the long side on the way up, you'd have done very well. You *would not* have gotten into trouble by selling point T1, or by buying points T2 or T3. You wouldn't have made any of these trades, *even though Trend rules learned previously would have justified such a play.* (We will discuss Directional Indicators which overrule Trend in the next chapter. They would have you *buying* at T1, and *selling* at T2 and T3, which are all at near perfect Fibonacci retracement points. Pardon me for digressing, but a quick look ahead is sometimes useful.) Now, back to the point.

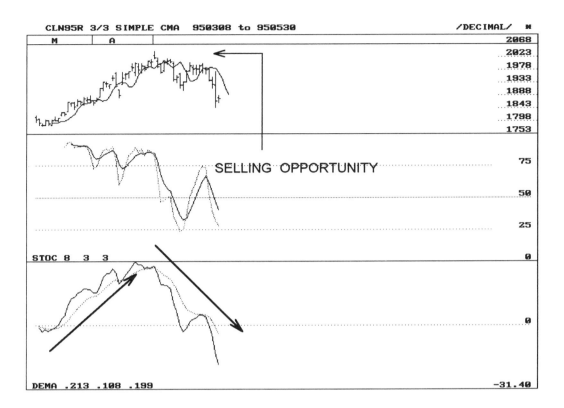

CHART 5-9

Chart 5-9 is the same daily crude oil chart depicted in Chart 5-8, with the MACD and Stochastic indicators added.

The entire downtrend after the high was contained by the MACD. The rally up to Fibonacci resistance at point T3, supported by the Stochastic, gave us a perfect opportunity to *get short*.

The same can be said for the up move preceding the high. The up move was almost totally contained by the MACD, while the Stochastic gave us ample opportunities to get long when the market retraced.

FREQUENTLY ASKED QUESTIONS:

Wouldn't it be better to have two strong Trend indicators, instead of one weak, and one strong ?

No. The weak Stochastic shows the hand of the weak players. It can also show the strength of the market. If the Stochastic gives a sell and there's no discernible movement down in price, watch out for a big move up!

Conservative Carl: *Do you wait for the bar (time period) to close before making a determination that the indicator has given a signal?*

This question leads us into level 3 understanding. By definition, you want confirmation to be certain of the signal. But, if you wait for confirmation, a significant part of the move may be in place. By waiting, you are paying for insurance you may not need. Just as it is acceptable to anticipate the DMA crossing of price, *before* the period closed, it is also acceptable to anticipate these Trend signals, *prior* to the close of the period. Be sure you get confirmation of what you have anticipated by the close, or get out immediately!

I've always bought when I got a Stochastic buy. How can I now sell?

If you are going to be among the 5% to 15% winners, you will have to be open to applying methods and procedures that are different from those of the masses. If it were as easy to win as following a Stochastic crossover, where would all the losers come from to pay the floor, as well as the off floor winners?

Hyper Hank: *So I buy when I get a sell on the Stochastic and sell when I get a buy on the Stochastic, right?*

No, you fade the Stochastic in a thrusting market in the context of a Trend, supported by the MACD. You don't simply *BUY* or *SELL*. Then you employ methods of entry, covered in CHAPTERS 8, 9, 10, 11 and 13.

Why don't you use the 25/75 barriers before assessing a buy or sell signal on the Stochastic?

Any crossing of the fast line through the slow line is considered a signal because of the unique way I am using the indicator. It may also be useful for you to note that trading experience (not formal computerized research) indicates that stronger signals, on both the MACD and Stochastic, are typically given if there is a greater angle of attack at the point of crossing. See Charts 5-10 A & B. The stronger appearance generally is indicative of markets that are moving and turning, rather than consolidating.

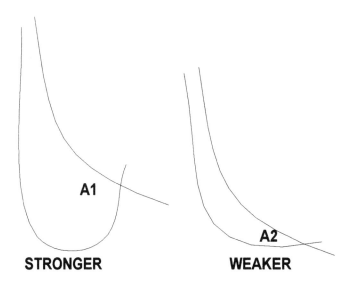

STRONGER   WEAKER

CHART 5-10 A & B

Diligent Dan: *"In the first example of daily bonds (Chart 5-5), it looked like the MACD barely gave a sell near the end of January and then got right back to the up side. Is there some way we could have avoided being whipped as we saw this action unfold?"*

Since this was a daily-based signal, it's likely I would have been whipped as you suggest. It's unlikely I would have wanted to take home a position that was against the MACD. There are, however, ways of avoiding being whipped under similar circumstances. If, for example, the Fibonacci support levels have *not* been penetrated by price, you could have a trading plan that would accommodate a minor break of the MACD. This would give the MACD an opportunity to correct . I give myself this latitude if the MACD signal is of the weaker variety, as in Chart 5-10B A minor break of the MACD on an *intraday* chart is also easier to live with than a break on the daily, since you can find out quickly if the position will hold on the anticipated Fibonacci retracement. You may choose not to allow this much discretion in your trading plan, until you gain more experience working with the concept.

SUMMARY:

Let's summarize the high points of our second Trend analysis tool, the MACD/Stochastic combination.

• Both the MACD and Stochastic give Trend signals when the fast line penetrates the slow line. These signals remain intact until another penetration occurs. The signal is *confirmed* at the close of the period.

• The MACD/Stochastic combination is applicable for all Time Frames we use.

• The values I use for the MACD (DEMA) are .213, .108, .199.

• The values I use for the deliberately weakened *Preferred* Stochastic are 8, 3, 3.

• To get good results using these indicators, you need to investigate the formulas used to create the studies involved as well as the method of programming the mathematical inputs into these formulas. The graphics software used to display the charts should show market-aligned bars, not time-aligned bars.

• By utilizing a specific weak Stochastic and a specific strong MACD, we are able to make informed judgments about what the weak and strong players are doing. Consequently we can determine how to best interact with price to achieve our goals.

CHAPTER **6**

DIRECTIONAL INDICATORS
9 POWER PATTERNS FOR HIGH PROBABILITY TRADING SIGNALS

GENERAL DISCUSSION:

If you don't understand or remember the definition of DIRECTION, FAILURE, or MOVEMENT, you'll make it a lot easier on yourself if you review these concepts in CHAPTER 2, before going on.

A Directional Indicator is usually, but not always, a pattern of some kind. Certain of these patterns are defined by the 3X3. Don't confuse the concept of *trend* with that of *direction*, even though the 3X3 is used for trend delineation as well as the criteria for certain of the Directional Indicators we will use. Directional Indicators are not Trend Indicators, irrespective of how they are derived. If there is a conflict between what a Trend Indicator and a Directional Indicator is telling you, follow the Directional Indicator. *Direction overrules Trend.*

In recent years, a greater percentage of my trades have been based on Directional Indicators rather than Trend Indicators. This is as much a reflection of my maturity as a trader, as it is my increased patience level. Perhaps philosophically speaking, those aspects are one and the same. Directional Indicators require patience, as they must manifest on their own, while a Trend can always be found. Directional Indicators are typically very powerful and highly reliable. They are in keeping with my overall philosophy:

LOSS OF OPPORTUNITY IS PREFERABLE TO LOSS OF CAPITAL!

THE "DOUBLE REPENETRATION SIGNAL" OR "DOUBLE REPO" FOR SHORT

The following Chart 6-1 shows an idealized Double RePo. It has specific, identifiable features.

1. The Double RePo signal bar must be preceded by a minimum of 8-10 periods of thrusting market action; 15 or more is better. Just what constitutes thrust is far easier to visualize than to define. This is good news, since it will make it tough for programmers to specify. Therefore, it should continue to "work" for some time to come.

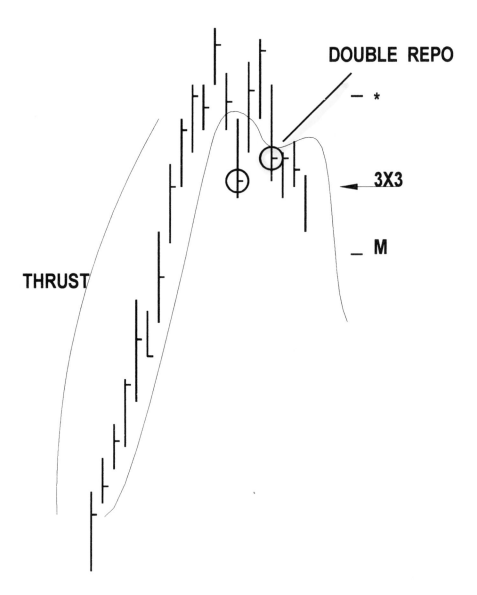

CHART 6-1

2. After the up thrust, we need closes below, above, and again below the 3X3, before the sell signal is given. The reverse is true for down thrust.

3. The top (or bottom) formed by the bars as specified in the chart shown, should be reasonably close to one another.

4. The width of the top (or bottom) from the initial penetration to the subsequent penetration (after the retracement) should not exceed 8-10 bars; three or four are much better.

5. The signal remains intact until either a major Logical Profit Objective (to be defined later) is achieved (point 'M' on the chart), or until the .618 retracement '*' from the furthermost extreme of the consolidation area (after the second penetration) to the furthermost extreme on the thrust, has been exceeded on *close*. When we get into Fibonacci analysis, the last sentence will be easier to understand. In terms to be defined later, you create a resistance series and look for the '*' retracement level to be exceeded *on close*.

6. The periods I use for Double RePos are daily, weekly, and monthly, although many of my clients have reported excellent results on 30 minute and hourly charts .

Let's review each of the criteria on this idealized chart, to be sure you understand the criteria, before we get to actual market examples.

ITEM 1
Market action is in a thrusting up move, remaining (primarily but not necessarily) in an up trend (above the 3X3 on close) for 13 periods.

ITEM 2
We finally achieve a close below, above, then below the 3X3, yielding the signal bar. Note the circled closes. The second of the two is the signal bar.

ITEM 3
The tops formed by the highs of the consolidation area are acceptably close to one another.

ITEM 4
The number of bars in between and including the two closings below the 3X3 are clearly less than the maximum.

ITEM 5

The signal would be negated if the market closed back above the Fibonacci level shown as '*', or if the market achieved the profit objective point 'M.' Neither situation has occurred yet.

IMPORTANT POINTS TO NOTE:

The criteria as defined in Items 1 through 5 will clearly satisfy many of you, and utterly disappoint some of you. The level of specificity as applied to the accompanying real charts should be more than adequate for those of you who are psychologically suited to judgmental trading techniques. I will endeavor to be as specific as possible with the Double RePo as with all of the subsequent Directional Indicators. It's important for you to note that just because the exact criteria are not met, *it does not mean* that the market will not *act* as if the criteria have been met. See "Look-Alikes" later in this chapter. In workshops, I sometimes refer to a Double RePo "look-alike" as one Double RePo not being as "pretty" as another. A Double RePo does not need to be a Barbie doll to add significant width to your wallet.

It's also important to note that a Double RePo on a weekly or monthly chart can be a serious event. It can signal the termination of a major bull or bear run.

Chart 6-1A shows typical market action after a Double RePo has occurred.

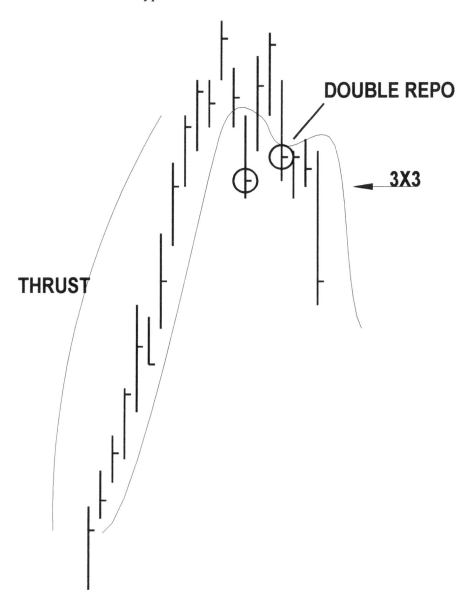

CHART 6-1A

FREQUENTLY ASKED QUESTIONS:

How did you come up with the signal?

Like *all* of the Directional Indicators in this chapter, the criteria were developed by my personal experiences in trading, some of which were blistering events! When dramatic price action occurred, I wanted to know why, and since I typically had my own, *real money* riding on the event, I was a highly motivated researcher. When I thought I learned something, I looked for the market to replicate the action again, and again, and again, and that's right, again, before I would consider adding what I thought would qualify as a directional signal to my trading arsenal. The standard of reliability was high, since my trading action subsequent to its identification was immediate and strong. The Double RePo signal was actually identified originally by me on the first chart shown in the series of actual market examples, Chart 6-2. It has been an incredible trading resource since 1986!

Can you anticipate the close on a Double RePo and take a position based on what might happen?

Yes. If, for example, the price on the signal bar shown on the idealized chart is below the 3X3 but the market hasn't closed, it is okay to get short. You could even anticipate the move through the 3X3 if your intraday trend signals, the MACD/Stochastic, were convincing enough. As with any anticipated or unconfirmed signal, however, you say adios to the trade if you lack confirmation at the close of the period.

Aren't you simply showing a double top?

No, I am defining a specific type of double top or bottom.

What markets does this signal work in?

All liquid markets. That includes stocks, mutual funds, and cash currencies. But, let's exclude wheat and pork bellies to be on the safe side.

Do I take a signal on close and simply wait for a profit objective?

You can take a position on a confirmed or unconfirmed signal, but you have to be aware of the "get out" point on the loss side as well as the anticipated profit. You can also apply all the other aspects of trade entry we have yet to discuss. One way to approach trade entry for multiple contract players would be to lay on some contracts unconfirmed, more on a confirmed signal, and still more according to other criteria we have yet to discuss.

What if we have a Double RePo and then close above the 3X3, but we do not exceed the Fibonacci point marked '' ?*

In the scenario you describe, the Double RePo signal has not been negated. Remember, you not only need to exceed the Fib point at '*', but you must also do so on *close* for a *confirmed* get out. As for the Trend confirmation up (confirmed by closing above the 3X3), you ignore it, since *Direction overrules Trend.*

What if you exceed point '' on close then turn around and subsequently close back below the 3X3?*

At this point, you're probably unhappy and counting some losses. Triple RePos don't exist in my trading plan in the overall "applicability" and "high reliability" sense that Double RePos do. However, I have observed a number of instances where such market action has occurred in U.S. bonds. The bond market has a nasty habit of doing this after very extreme thrusting moves. I trade such action similarly to a Double RePo. This would be classified as a "look-alike."

Why do you limit the periods you use to daily and longer ?

They are the most reliable. A number of my clients use intraday charts for the Double RePo trade as well. If you want to see what one looks like, there's an example of a Double RePo on a 30 minute S&P in CHAPTER 4, Chart 4-5.

Can you explain what is actually happening in the market that makes this signal work?

I can try. The day after day of thrusting moves demoralizes and panics the shorts. Longs that have exited are pulling their hair out from feelings of greed. Most lack the ability to get back on board. The first pull back is bought by these groups, while the second pullback eventually turns into capitulation or even panic, if the preceding up move is confined by the '*' Fibonacci barrier. The important point of this psychological discussion of what's happening is understanding that the length of the initial thrust must *not* be unduly consolidated by the width between the 1st and 2nd penetration of the 3X3. In other words, 18 days of up thrust consolidated by six days is a lot "prettier" than eight days of thrust consolidated by eight days. Too much consolidation works off the greed and fear. We don't want that.

MARKET EXAMPLES:

Each of the following charts will attempt to identify and clarify the Double RePo Directional (change) Indicator.

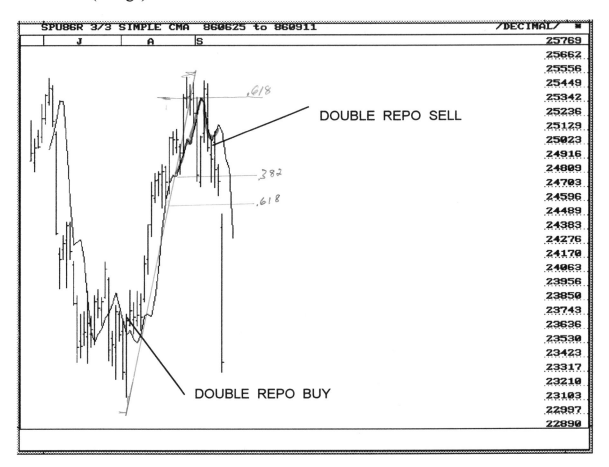

CHART 6-2

Chart 6-2 shows two Double RePo signals in the S&P, occurring in 1986. I had my proverbial lunch handed to me just before the first signal. The locals in the S&P had their lunch handed to them just after the second signal. We both learned something from the experience. I learned *about* the Double RePo signal. The locals learned to keep their hands in their pockets when the S&P freight train barreled through.

The second signal (on the sell side) is picture perfect. It was preceded by beautiful thrust, had near equal tops, and the span between the first and second close below the 3X3 was narrow.

The first signal (on the buy side) certainly worked. The down thrust preceding it however, was strong, but not *as relentless* as the up thrust that preceded the second signal. What we prefer to see in terms of thrust is continued pressure, as in the idealized example, rather than a single big move, consolidation, and another big move. While the levels of the double bottom were acceptably close, they were not as "pretty" as the double tops preceding the sell side signal. The span of days between the first and second closing above the 3X3 was also a bit wide, given the extent and nature of the preceding down thrust.

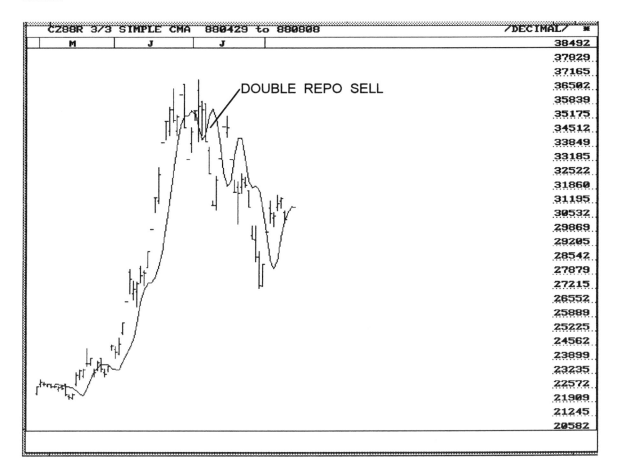

CHART 6-3

Chart 6-3 illustrates a perfect Double RePo right at the top of the weather market we experienced in corn in 1988. For those of you who haven't been involved with weather markets, they are among the most *vicious* of *all* markets. The dots on this chart do not indicate illiquidity brought about by lack of interest; they are limit moves!

The next Chart, 6-4, shows weekly continuation data of corn, preceding and including the crop shortage period in 1996. Note the clear Double RePo occurring near the $5.00 level. How would you have liked to have been short a ten lot when that happened?

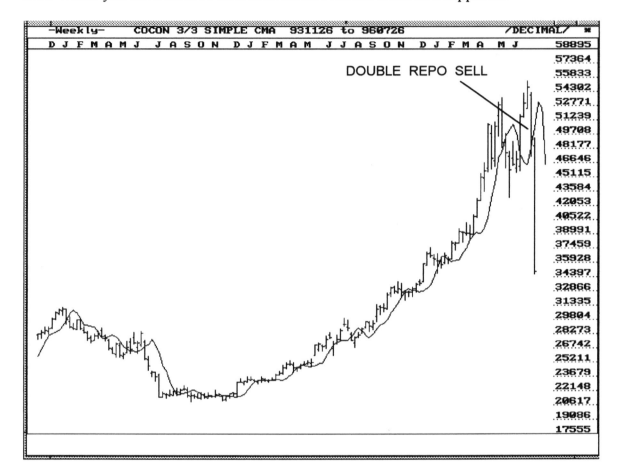

CHART 6-4

Realize that if you had a position on, it would have been in September Corn, not on the continuation chart. The Double RePo showed up there as well. The continuation chart was necessary, however, because it presented you with a cleaner picture. This thinking is similar to that used in an upcoming soybean meal trade, detailed in CHAPTER 15. In the meal trade, I entered on July meal but the context was taken from the weekly continuation chart.

Next, we'll see daily soybeans (Chart 6-5) during and after the floods in 1993.

CHART 6-5

Just for a change of pace, let's see how this Directional Indicator works in the stock market.

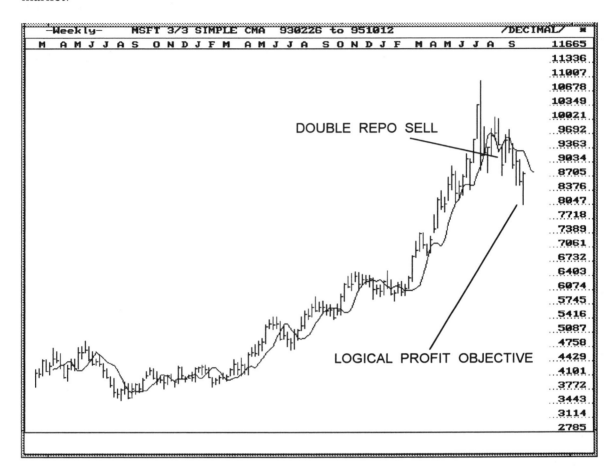

CHART 6-6

Microsoft (Chart 6-6) peaked at the introduction of Windows® 95 and made an almost perfect Double RePo. From there, it fell to a precalculated Fibonacci profit objective. Not only was this an opportunity to book weekly-based profits on the sell side, it was also a time to get long on a monthly reaction to an ongoing up trend. It would not serve you to skip ahead now, but some very interesting charts on Microsoft in the monthly time frame are just ahead. For those of you who are still having trouble internalizing the concepts of Time Frame, Trend, and Direction, these upcoming charts should be very helpful.

How about crude oil? Well, it's certainly a liquid market. It's also a volatile market, and that's what we want. Let's see what happened when Saddam did his thing back in the summer of 1990.

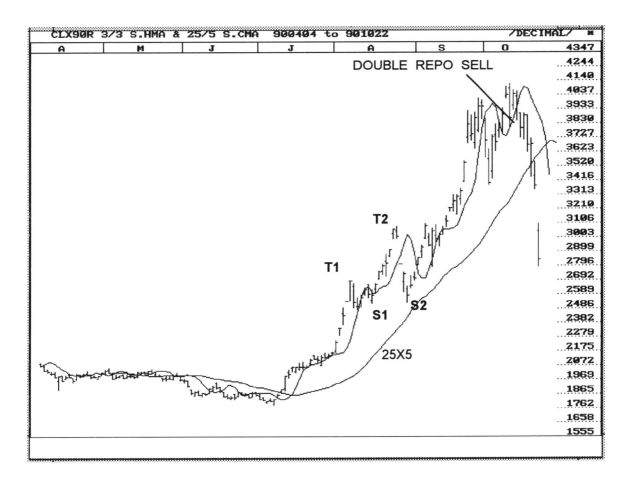

CHART 6-7

Chart 6-7 shows the 3X3 containing trend all the way up to the $40 level, then a Double RePo and subsequent break through the 25X5. This Double RePo was the first time since the invasion of Kuwait that it was "safe" to go short. The previous breaks of the 3X3 (S1 & S2) were single penetrations only, primarily due to the *vast difference* between the two tops (T1 & T2) made prior to each penetration. These dips presented buying opportunities as you will see when we cover our *Bread and Butter* directional signal.

If you had misinterpreted these single penetrations as a Double RePo, you would have shortly reversed and more than made your money back when the '*' Fibonacci level (not shown) was exceeded. This is called a Double RePo Failure. It will be the next directional signal we cover.

In Chart 6-8, we have a monthly based, Double RePo buy in the German bund.

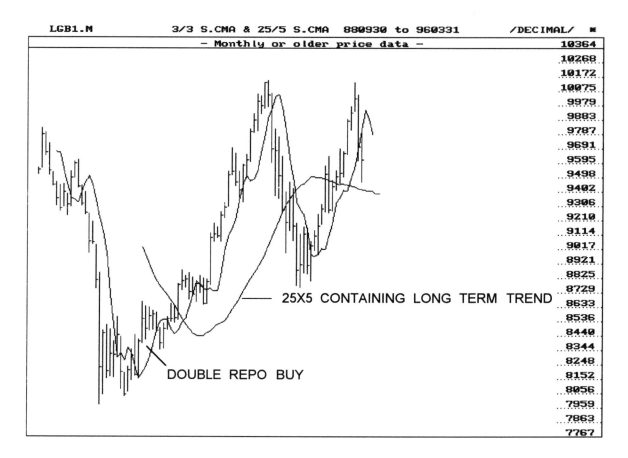

CHART 6-8

I could go on and on with these examples, but you should have enough here to recognize this Directional Indicator when you see it. I've stressed your understanding of the Double RePo since your understanding of its formation is critical to your understanding of the Double RePo Failure as well as Bread & Butter. Before we leave this topic, however, I want to show you a very interesting Chart (6-9) of monthly gold.

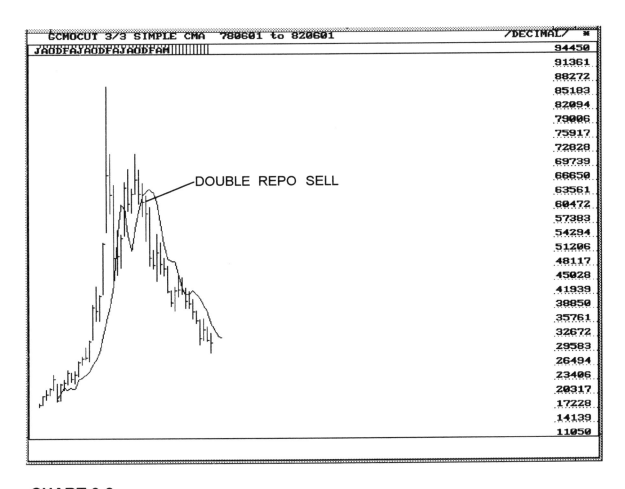

CHART 6-9

I didn't know about the Double RePo back in the early 80s when this event took place. I wish I had.

The original move up to 875 certainly qualified as thrust. The dissimilar tops were a bit hard to live with but the fact that the second peak was held back by Fibonacci '*' resistance made it more palatable. The sell signal was given at about 625. The subsequent fall was a move down to about 280! I'm certainly *not suggesting* that anyone trade this monthly chart with what might be a $200 stop. What you need to understand is that the monthly set up can be used for weekly-based mutual fund switching or daily-based commodity trading. Another reason I wanted to show you this chart is so you can see what I'm looking for in the current stock market rally *before* I get excessively bearish. *If* we get a monthly Double RePo sell on the Dow or in the S&P, say good-bye to stocks! If the ensuing monthly pullback returns to the .618 retracement from the beginning of this bull market, we could be looking at a loss of over 4000 points!. If we only achieve the .382 retracement, we're still looking at something over 2500 points! For those of you who don't think this is likely or at least possible, gaze at Chart 6-9 again.

THE "DOUBLE REPO FAILURE":

1. First, you must have a Confirmed Double RePo, as shown in Chart 6-10.

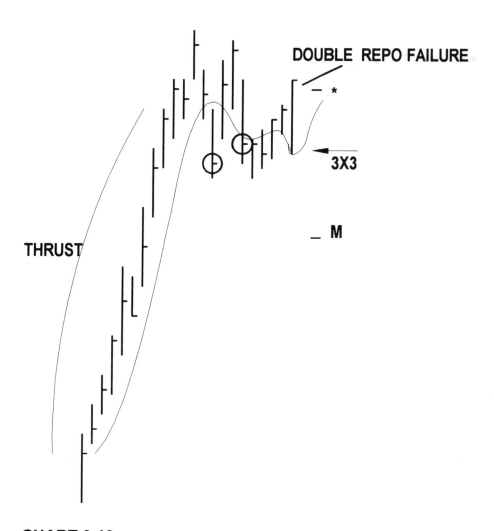

CHART 6-10

2. The Double RePo signal fails and is negated when the closing price exceeds the Fibonacci '*' level. This is your signal bar. The expected subsequent action should be strongly up.

3. Your exit is at a significant Logical Profit Objective or a *Confirmed* Trend signal as defined by the 3X3 that does *not* confirm the action expected from the Failure.

Note: This signal is one of the few instances when you would reverse your original position. It would also be acceptable to aggressively enter (initiate a new position) against the high of the Double RePo Failure signal bar, after even the smallest of pullbacks. The point is, this is a Directional signal. You don't play with it. You get out of the way and go with it! In Chart 6-10, we clearly have a Double RePo which fails by exceeding Point '*'. Once you're in the trade to the long side, you have two ways to exit. Point COP (not shown) would be your Fibonacci precalculated, profit objective. Penetrating back below the 3X3 on close is your protective exit. This protective exit could turn out to give you a profit or a loss on the trade, depending on when and where price action crossed the 3X3. You could also use the stop placement methods discussed in CHAPTERS 8 through 11. Regardless of alternative tactics, the *signal stays in play* until a significant Logical Profit Objective is achieved or you get a Confirmed Trend against the anticipated Movement.

FREQUENTLY ASKED QUESTIONS:

Once the Failure is in place, how aggressively do I treat it?

You immediately get out of any existing positions that are counter to the action anticipated by the Failure. You may initiate new positions aggressively, or by criteria yet to be described (dropping the Time Frame and entering at a Fib retracement or Confluence area).

Can I use the MACD/Stochastic Trend indicator to determine the protective exit rather than the 3X3, if the Failure doesn't go my way?

Yes, a Failure should move now, and keep moving. If it doesn't, something is wrong.

Chart 6-11 weekly bonds has two Double RePo Failures as well as a Double RePo. I've identified one Double RePo Failure and the Double RePo. Can you find the other Failure?

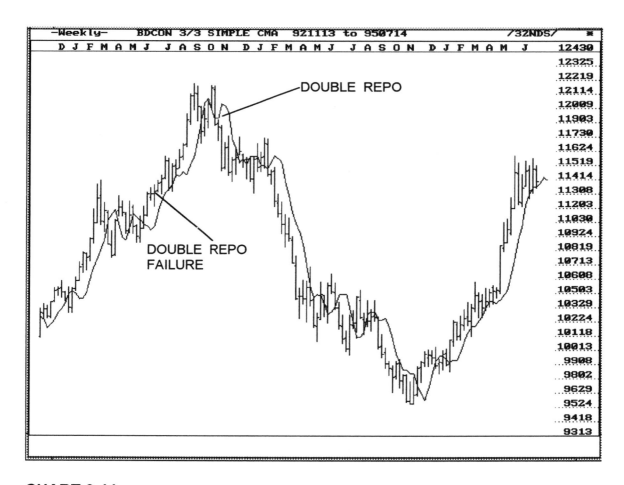

CHART 6-11

Chart 6-12 shows a Double RePo Failure in the daily bund. The price action to the right of the top does not qualify as a Double RePo, since there was not adequate thrust. Note how nicely the 25X5 contains long term Trend.

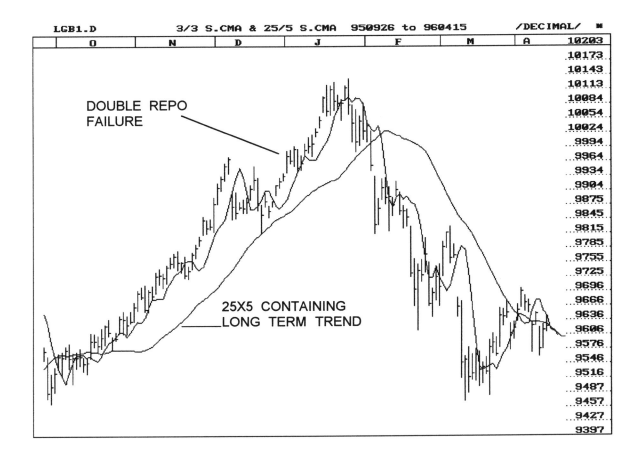

CHART 6-12

THE "SINGLE PENETRATION" OR THE "BREAD AND BUTTER" SIGNAL:

I'd like to say this is a "lay down" but of course nothing is in trading. This trade is designed to book you some nice comfortable profits with little risk. Unfortunately, it requires a good understanding of *advanced* Fibonacci Retracement Analysis to implement, so you will need to reread this after D-Levels™ are covered. Some of the terms, of necessity, must be broad.

1. Like the Double RePo, the Bread and Butter must have a minimum of 8-10 periods of thrusting market action. More is better. Just what constitutes thrust is far easier to visualize than to define. This is good news, since it will make it tough for programmers to specify. Therefore, it should continue to "work" for some time to come.

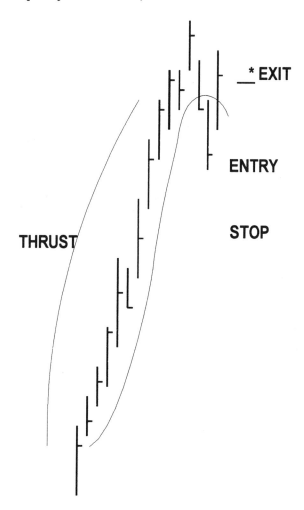

CHART 6-13

2. After the initial penetration of the 3X3 on close, look for an intraday Fibonacci (support) retracement level, at a significant Fibnode, to enter the market *in the Direction of the original thrust*. This level should occur within one to three periods of the initial Confirmed break of the 3X3. I recommend daily, weekly, and monthly periods, although this strategy will also have merit on intraday charts. The retracement Fibnodes which are the basis of your entry as well as your stop, should be calculated from the hourly (and higher) Time Frame charts, if you use the periods I suggest.

3. Once the trade is entered, set your stop loss beyond a deeper Fibonacci retracement level, and your profit objective a bit before the 618 retracement '*' of the entire contra move, i.e. the move that is opposite the original thrust.

Let's take a look at the monthly gold Chart 6-14, as an example of our buy and sell points.

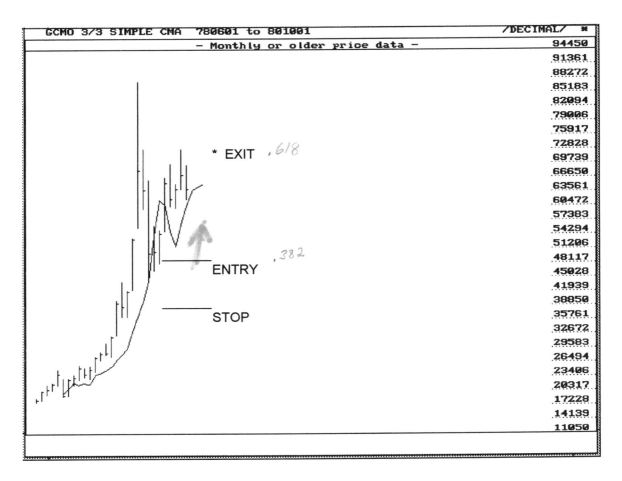

CHART 6-14

Before you have a coronary over the prospect of that $200 stop, let me point out that this is a theoretical example. It could be made practical, however, if we used the monthly Direction for weekly-based mutual fund switching or even daily based Fib entry. Context! Context! Context! Plan your trades as you would any other important financial endeavor. I'd give you more examples, but we haven't covered how to arrive at the Fibonacci entry and stop levels yet, so the exercise would have limited relevance.

Before we leave this topic, however, let's look at our friend Microsoft, Chart 6-15. This time we'll view it in a monthly Time Frame.

There's no inconsistency in being a seller on a weekly Double RePo, as discussed earlier, closing the trade at the Logical Profit Objective shown in Chart 6-6, and then being a buyer based on a monthly Bread and Butter signal. If this seems inconsistent to you, you don't fully understand price vs. time charts or Time Frames. A review of the Trend discussion in CHAPTER 2 may help.

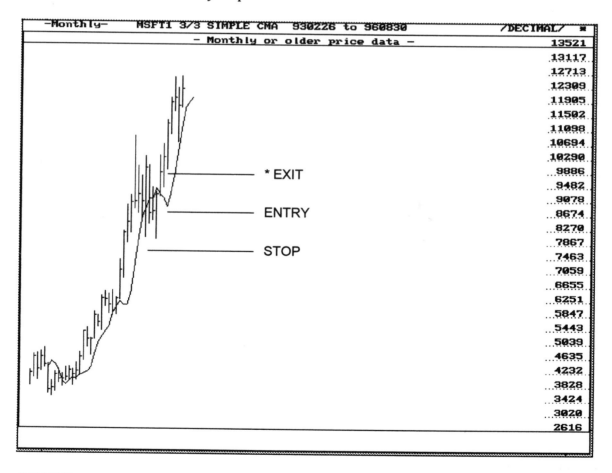

CHART 6-15

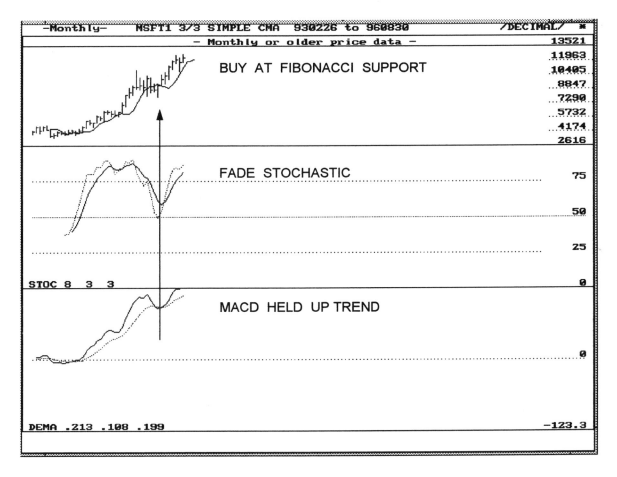

CHART 6-16

Notice how the monthly MACD Confirmed the ongoing monthly up trend, while the Stochastic gave us a great buying opportunity, once the profit objective shown in Chart 6-6 was reached.

Note the consistency between the Bread and Butter monthly buy and the MACD shown above. What is not shown is a daily Double RePo Look-alike on the buy side that occurred at the same time! *This is what high probability trading is all about!*

PATTERN FAILURES:

The idea behind these directional signals is *not* to follow everyone else, but rather to fade those pattern players when you are sure they are wrong. Typically, newer traders are looking at the standard Robert Edwards and Magee trading patterns[1] for trading signals. They make an excellent group to feed off, *if* you can determine what they are doing and when they are likely to panic. These signals work best when these traders have the time to get in the trade the wrong way. That's why I look at daily, weekly, and monthly pattern Failures. While the higher Time Frames hold the more secure probabilities, some pretty dramatic results can be achieved from certain intraday pattern Failures.

THE "HEAD AND SHOULDER FAILURE":

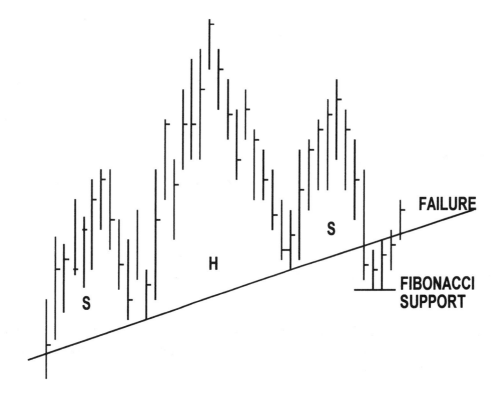

CHART 6-17

The above idealized example illustrates a clear H&S with a break of the neck line, a bit of consolidation below it, then a close back above it. The subsequent action is expected to be strongly up because the short selling pattern players are caught wrong, and must

---

[1] Robert Edwards and John Magee, *Technical Analysis of Stock Trends*

unwind their positions. The lower portion of the consolidation after the break of the neck line may be supported by a significant Fibonacci level. If it is, you will have some advance warning that a Failure is coming and you can enter according to the Fibonacci tactics taught in CHAPTER 13. It isn't necessary however, that such support causes the subsequent Failure. *What's necessary is that the Failure occurs.* Anticipate this pattern at your risk. If you anticipate this signal before it crosses the neck line, you will be trading against a classic pattern and you would also likely be against the prevailing trend. Remember, you are *not* taking the classic sell signal, rather the Failure (buy), *if* it happens.

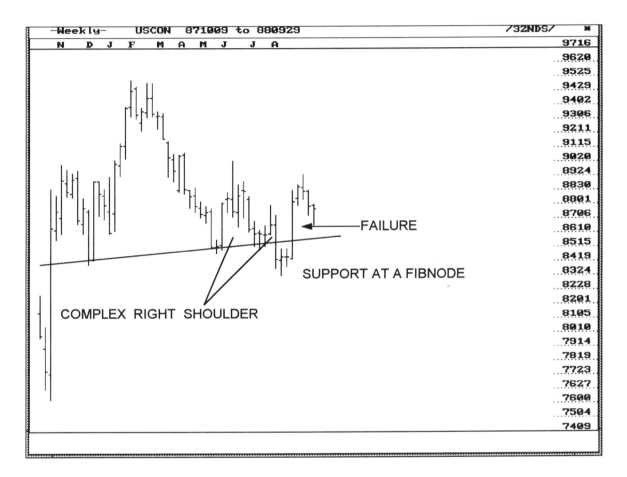

CHART 6-18

The bond weekly Chart 6-18 shows a picture perfect example of this phenomenon. We have strong Fib support below the neck line. Time for players to "get wrong" (two to three weeks) followed by a subsequent sharp move up, trapping those pattern players. This weekly action would be the set up for the trade. You would enter on the daily Time Frame.

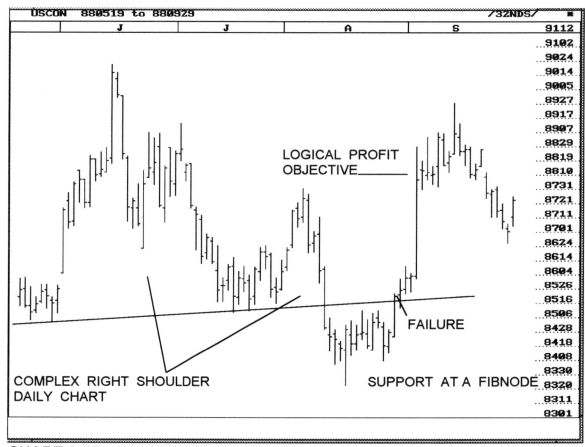

CHART 6-19

The daily chart illustrates how powerful and rewarding this signal can be. The idea is to drop your Time Frame to enter the trade, once the setup is apparent.

If you are observing the pattern Failure on a weekly chart, you can enter on the daily. If you are observing the phenomenon on the daily, enter on the hourly. The specific entry techniques you would use will be covered later in this book. If you hesitate, you can easily get left behind!

NOTE: These types of pattern plays are particularly rewarding if they are widely promoted in a specific market, particularly on TV programs, or in a widely-followed newsletter, or fax service.

THE "TRIANGLE BREAKOUT FAILURE" OR "OOPS":

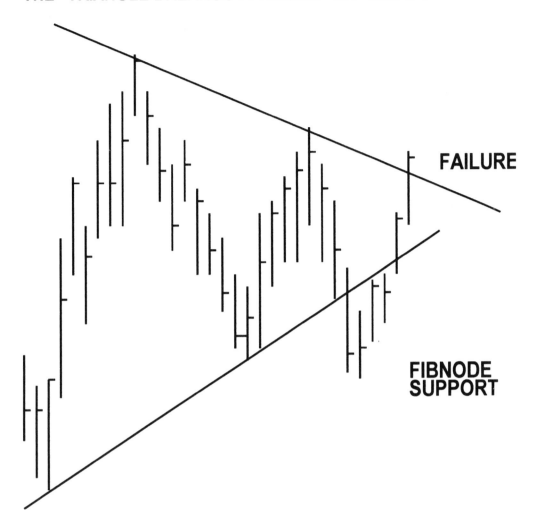

FAILURE

FIBNODE
SUPPORT

CHART 6-20

The Triangle Breakout Failure or OOPS can take a variety of forms. The key is that traders have *recognized* that the triangle pattern exists and they have had *time* to act on it, thereby getting in the wrong way. It can't be too subtle. All the same reasoning applies as with the Head & Shoulders Failure.

"FADING POPULARITY" OR "THE VULTURES DELIGHT":

In the early 90s, when Candlestick charting was new in the U.S. and aggressively promoted, I had a studious client call me whenever the strongest of the Candlestick signals occurred. I never took any of the signals. I simply asked him, as a dedicated devotee of Candlestick patterns, when *his* Directional signal would indicate that he was wrong. That was *my* Directional signal. Soon he understood what I had been trying to teach him about how powerful Failures can be.

THE "RAILROAD TRACK":

This is a top notch Directional signal that can occur *in any Time Frame*. Its wide applicability is likely responsible for the glowing comments from my clients. It's as easy as it gets to apply and profit from, with a minimum of effort. For those of you who have studied Steidlmayer's work [2], the Market Profile®, and understand "rejection of price," the underlying concept of why this works will be apparent. To those of you who are unfamiliar with these concepts, just imagine a half a dozen ocean-front homes in Santa Barbara, California suddenly going up for sale at $100,000 each. Boom, they're gone. Professional investors and Realtors snap them up and the price is back to the more normal range, perhaps higher since *overhanging supply is now absent*.

The idealized Chart 6-21 depicts a typical Railroad Track. With the extensions down as shown, we would expect a strong up move. Notice how the extended two bars are out of close proximity with the others bars. We call this "Railroad Tracks in the country", i.e. nice, scenic, and pleasant. Railroad tracks in the city are congested, unhappy events, potentially dangerous to cross, smoggy, and unsightly. As futures traders, what we want is space around the Railroad Tracks we select to trade. For rejection of price, we need price that is not common.

The Railroad Track (RRT) can take place in any Time Frame, from five minute to yearly. It is one of two directional signals which allows you to adjust the Time Frame to any desired amount. In other words, you may cause the data to fit into our perception of usability, by adjusting the Time Frame. So, let's imagine some variations that will all work, since they all represent the same phenomenon.

---

[2] J. Peter Steidlmayer and Kevin Koy, *Markets & Market Logic*, (The Porcupine Press, 1986)

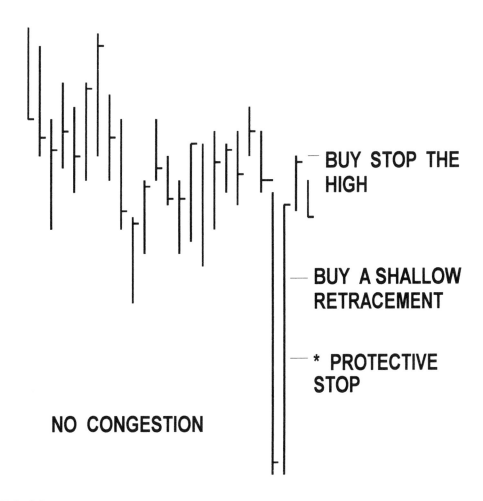

BUY STOP THE HIGH

BUY A SHALLOW RETRACEMENT

__* PROTECTIVE STOP__

NO CONGESTION

CHART 6-21

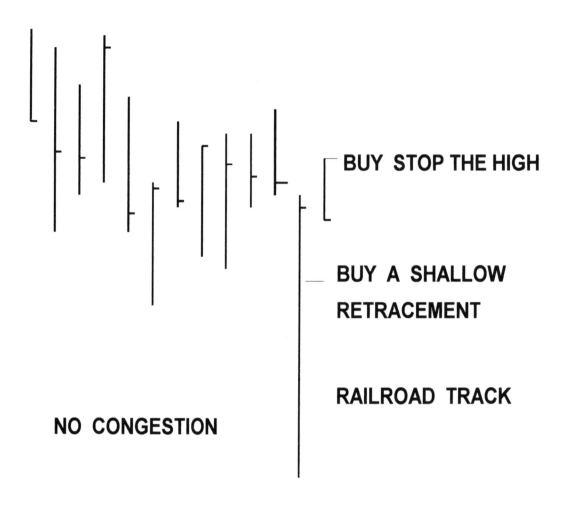

BUY STOP THE HIGH

BUY A SHALLOW
RETRACEMENT

RAILROAD TRACK

NO CONGESTION

CHART 6-22 (TWICE THE TIME FRAME)

Chart 6-22 contains exactly the same data as Chart 6-21, only it is shown at twice the Time Frame. If Chart 6-21 was a half hour chart, Chart 6-22 would be an hourly chart. If Chart 6-21 was an hourly, 6-22 would be a two hour chart. The RRT phenomenon works exactly the same, since the price action is exactly the same. It just looks a little different.

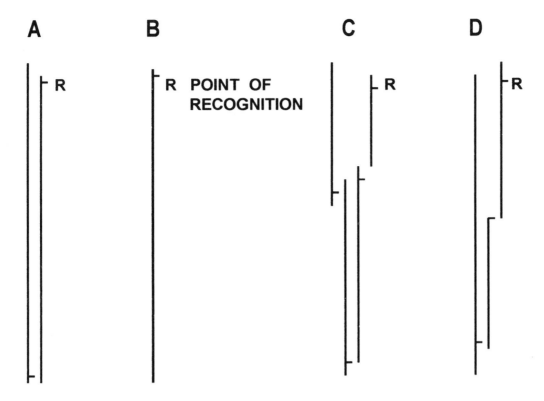

A   B   C   D

R   R POINT OF   R   R
              RECOGNITION

RAILROAD TRACKS IN DIFFERENT TIME FRAMES

CHART 6-23

Chart 6-23 shows a variety of looks that RRTs can assume. You'll need to get used to the fact that all of these forms are applicable in exactly the same way, since they depict the same phenomenon.

So, how do we play it?

These events can be so powerful that some of you will choose to enter "at the market" once the RRT is established, while others will choose a Fib retracement level for entry, as described later in this book. You will know you are wrong if the RRT returns to its extreme in price. It would be advisable to have a stop in excess of the .618 retracement, the '*' Fibnode. It would be nice if you could enter on a .382 retracement of the move up, after point R (recognition) is achieved, but often the continuation movement is too powerful. You will likely need to employ the more advanced methods covered later. I will typically stop myself in (initiating positions) against any extension of the extreme of the move *after point R*, while also attempting to position myself on a shallow retracement level. If I'm filled on both orders that's fine. If I just get the initiating buy stop, that's okay too.

In the following 30 minute chart of U.S. Treasury bond futures, we can see a Railroad Track extending up to 11525. We know it's a RRT when subsequently the market returns to where the extension started, approximately 11508. That's the point of recognition (R). Note the flat tops at 11514. They show a large overhead supply. This is our sell point. We can also initiate sell stops at the 11508 level in case we're not filled at the retracement sell, or if we want to double up. This is a 30 minute chart. We don't want to be married to the position. We take a quick Fibonacci profit level (OP) near the low at 11428. If the higher Time Frame trends (not shown) were in a sell mode, we might want to stick around for bigger profits, or possibly hit it again on a retracement back toward the extension high. Based on the intraday RRT shown however, this was a nice "safe" trade that allowed us to book some nice "safe" profits. I like taking profits. So should you.

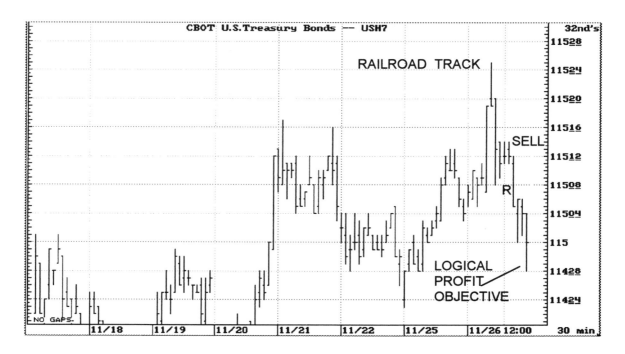

CHART 6-24

The corn futures chart below shows two RRTs. It was constructed outside of our "normal" Time Frames (see CHAPTER 2) to show you that we can "play" with the Time Frame selection to achieve the look of a RRT if it is not immediately apparent. This is what I mean when I say we can act on this signal in *any* Time Frame.

Note how the close of the second bar of RRT 1 bounced back from the extreme low, the point of recognition. Buried in the lower Time Frame action is the retracement we're looking for as our short entry.

RRT 2 was effective, but the contract was so close to expiration we'd have to look at the continuation chart or distant contracts for confirmation of this signal. Unless we were a commercial acting in the cash market, we would obviously trade on the signal in a more distant month than the one shown on this chart.

CHART 6-25

Like many of the Directional Indicators I've given you, I'm aggressive about my entry technique when the pattern develops. It is very dangerous, a prohibition in fact, to anticipate this particular signal. The reasons will be clear when we cover the more advanced Fibonacci techniques.

As a point of interest, I learned about this signal back in 1988 when I was long the S&P. A rumor about Vice President George Bush, fooling around on Barbara hit the market.

When the rumor hit, there was a hard break in price. It drove all the intraday trends down, albeit Unconfirmed (both the Trend and the rumor.) I hit the retracement back, i.e. liquidated my longs and went net short. The next thing I knew we were back to the top of the RRT. Now I was looking at two losses. I got stubborn and refused to take a short retracement back in the direction of my sell to get flat. Even though the Unconfirmed down trends disappeared, I was now trading against the Confirmed up trends I had originally used for the context of my initial long trade! This was clearly a *Mistake*. I wasn't going to let those (expletive deleted) get me! Well, they got me and a half an hour later when I finally got out, I was obviously counting more losses than if I had exited when I should have. Fortunately, I straightened things out quickly the following morning by *buying* the first retracement against a shallow Fibnode. The RRT was born!

I've included one final Chart 6-26 on gold. If you can't find the RRTs, you'd best see an optometrist.

CHART 6-26

"LOOK-ALIKES":

Look-alikes are "want-to-be's" or "near miss" Directional signals. They don't quite meet the qualifications. Maybe a Double RePo is lacking in the extent or nature of the desired thrust. Possibly the tops between the first and second penetrations are a bit wider than the specified criteria. Perhaps the extension bars of the Railroad Track are a bit short or maybe the Railroad Track is occurring in "suburbia" rather than "in the country." The consolidation below a Head and Shoulders neckline might be one period rather than the several periods we'd like to see, before the Failure occurs. All these situations are Look-alikes. Often Look-alikes will behave similarly to the Directional signals they mimic, but not with the probability of success we would expect from a fully qualified Directional Indicator. I'll play these signals with less gusto, fewer contracts on the line. You might choose to let them go by. Either way is acceptable, just be sure you don't get in their way.

"STRETCH":

This Directional Indicator combines an oversold or overbought oscillator with strong Fibonacci support or resistance. It is obviously pretty difficult to describe this signal since we have yet to cover either concept. See CHAPTERS 7, 8, & 9.

The basic idea behind this indicator is to act on the *combined strength* of these two powerful, *differently derived* Leading Indicators, which indicate support or resistance when their respective price values are in close proximity.

Even though your entry is strictly placed, Stretch is more risky than other Directional signals. This signal has you buying down thrust and selling up thrust. Understand its implications before deciding to use it, place a physical stop (Bonsai or Bushes, Fib Tactics, CHAPTER 13), or monitor the position closely.

THE "FIB SQUAT":

This Directional Indicator is similar to Stretch in that you need strong Fibonacci support at a point where something else occurs. In this case, the something else is a Squat. So what's a Squat? As developed by friend and colleague, Bill Williams[3], the Squat is a function of the range of a given price bar and the volume, or TIC volume, that occurs while that range is being created. The basic idea is that high volume and little price

---

[3] Bill Williams, *Profitunity Trading Group*, 2300 Pilgrim Estates Dr., Texas City, TX 77590

movement indicate substantial support or resistance. Some of you old timers will remember the term "churning." That was a phenomenon in the stock market which occurred when you'd get tons of volume (60,000 to 80,000 shares in a day) and little price movement. The idea behind my approach to this indicator is to first look for likely Fibonacci support and resistance, and then see if a Squat manifests when that point is reached. If so, the Squat confirms the Fib support or resistance, and it's okay to act without a trend in your favor to support the trade. This indicator, like Stretch, is a more risky (lower probability of winning) trade than the other Directional Indicators I've outlined. It's good for the hyper types or over-traders among you, since you can use it all the way down to a five minute chart. Both Stretch and the Fib Squat yield a reasonable probability of a successful outcome.

For additional safety, the Fib Squat is most useful in a context, within a context. Let's say, for example, that you have a strong weekly trend in place and you are experiencing a pullback via a Stochastic sell on the weekly. The MACD is holding nicely. With several daily Fibnode locations to choose from to go long, it is helpful to see a Squat manifest at one of them, before jumping in.

SQUAT = greater (TIC) volume and a smaller MFI than the previous bar has.

$$MFI = rb / v$$

where:
rb is range of the bar in tics or points
v is TIC volume

Squats look like high volume, low range bars, and are fairly easy to see without concocting elaborate formulas. See Chart 6-27.

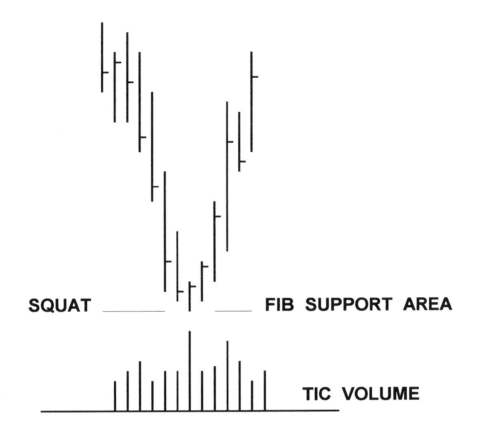

CHART 6-27

Below, Chart 6-28 is an equity on Fib support (not shown) experiencing high volume as well as a narrow range daily bar. The subsequent move up culminates at an almost exact Fibonacci Logical Profit Objective. After you complete CHAPTERS 8 & 9, come back and check for an OP move up. Incidentally, this up move turned out to be a very significant rally high.

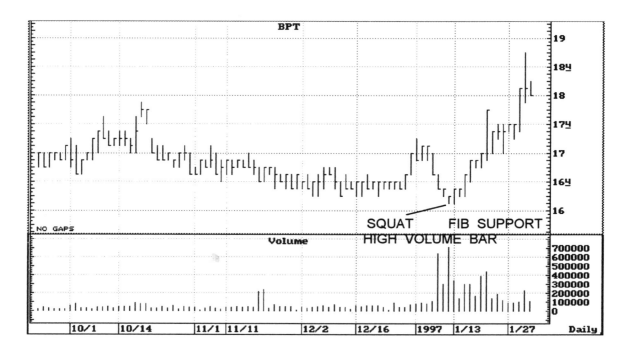

CHART 6-28

FILTERING THE "FIB SQUAT":

I have found that the level of "squatiness" is important as a filter to the Squat. If you choose to filter the Squat, part of the above formula would change accordingly: Squat equals 30% greater (TIC) volume and a smaller MFI than the previous bar. Larry Ehrhart, developer of WINdoTRADEr™[4], has done some valuable research in this area. The bottom line is that I like to see a significantly increased level of volume, say 30% or more, before I call a Squat significant. In addition, you may more easily see a Squat manifest in a four or six minute chart, rather than a five or 30 minute chart. As with the RRT, a Time Frame outside of the ones we normally consider is not only acceptable but desirable if it can help you identify the existence of the indicator.

---

[4] Larry Ehrhart, 3700 North Lake Shore Drive, Suite 7-09. Chicago, IL 60613, (312) 871-4687, (312)789-7434 Fax

You can anticipate a Squat if you have all the volume of the previous bar, a very tight range, and if only 1/2 or 1/3 of the Time Frame you are observing has elapsed. It's a pretty sure bet you'll get your necessary volume reading by the close, just make sure the range has not expanded unacceptably.

Another way to "play" the Squat, is to wait for a Confirmed Squat to occur in the vicinity of a major Fibnode, then enter on the first shallow Fibonacci retracement you get. Your stop goes at or just under the bottom (top) of the Squat bar. A similar technique is described as Minesweeper A in the section on Fib tactics, CHAPTER 13.

FREQUENTLY ASKED QUESTIONS:

Can I anticipate Directional signals?

You can anticipate any signal I give you in this book, but Directional signals can be among the most dangerous to anticipate. Anticipating classic pattern Failures and Railroad Tracks can be suicidal.

Why do Directional signals take precedence over Trend signals?

By their nature, they are more powerful.

A majority of your examples are for long plays. Are these signals as effective on the short side?

Yes, perhaps more effective since the public or less experienced traders have a propensity to favor longs. When they are wrong they are more easily panicked.

Don't you ever feel for the player on the other side of the trade?

It's your choice. Be the minnow or the shark. Minnows will be eaten and smart sharks get out of the vicinity when the big whites are circling.

In summary, I would encourage you to reflect upon what you see occurring in the market. Test the validity of reoccurring patterns and Failures of widely followed signals. Soon you will develop some Directional signals of your own. The difficult part is waiting for them to manifest.

CHAPTER 7

OVERBOUGHT & OVERSOLD OSCILLATORS
WHAT WORKS, WHAT DOESN'T, AND WHY
oooooooooooooooooooooooooooooooooooooooooooooooooooooooooooooooooooooo

GENERAL DISCUSSION:

Overbought (OB) and Oversold (OS) are among the least understood market conditions that traders grapple with. Most lose money attempting to utilize what they know about the subject. This is not surprising, because we're getting into the use of coincident and Leading Indicators and very few traders are properly prepared for the challenge these concepts present. Because of the level of misunderstanding, instead of narrowly defining what I use and how I use it, my approach will be to discuss the broad topic of Oscillators in general: what works, what doesn't, and why.

Typical thinking about Oscillators can be summed up by the following comment. "Oscillators *work* in a consolidating market, but once a Trend starts, they don't *work* at all." While this thinking may be typical, it severely limits and distorts a wealth of important trading strategies. The idea behind the statement goes something like this. You can sell Overbought and buy Oversold, as long as the market is consolidating and... you would expect to make money. This "sound good, feel good" strategy implies that you can identify *when* a market is consolidating with enough certainty to place orders. Does anyone want to try the ADX Average Directional Movement Index[1] as a means of making this judgment? While this approach may be acceptable for some of you, it's not for me. I

---

[1] J. Welles Wilder Jr., *New Concepts in Technical Trading Systems* (Trend Research, 1978), hereafter cited parenthetically in the text as Wilder, *New Concepts.*

haven't found it to be sufficiently accurate in this context, particularly for intraday charts. What about the second part of this statement?

"Once a Trend starts, they (Oscillators) don't *work* at all." The idea here is that the initiation of orders against the prevailing Trend will likely result in losses, by your stops getting hit.

The real problem with the initial statement is inherent in the way *works* is defined. I intend to show you how the right Oscillator can be made to *work* for you, in a wide variety of market situations.

Before we discuss uses and potential benefits however, let's first discuss which Oscillators are most commonly used, and which are *best to use* in the context of an Overbought/Oversold indicator.

THE STOCHASTIC:

The Stochastic is one of the most consistently misused indicators in the trader's arsenal. Traders typically consider any move over 75 to be Overbought and any move under 25 to be Oversold. That's not what George Lane (its originator) teaches and that's exactly the opposite of what Jake Bernstein's research of the Stochastic Pop Indicator[2] tells us. In fact, according to Jake's research, fully 50% of a strong market move can take place *after* the 75/25% barriers have been crossed!

In the daily treasury bond Chart 7-1, there are two places, marked by vertical lines, where you would have been *killed* buying in excess of 25% Oversold. Note, this is the case even though I have used the more typical (stronger) 14 period Stochastic, rather than what I described to you in CHAPTER 5. Further complicating the issue for new traders is the fact that in strongly trending markets, the Stochastic may never see these extreme (75%/25%) levels on typical retracements of an ongoing Trend. If you were waiting for these levels, you'd likely never get the opportunity to sell in a strong down trend or buy in an strong up trend.

---

[2] Jake Bernstein, *Short Term Trading in Commodity Futures*, (Probus Publishing Company, 1987), hereafter cited parenthetically in the text as Bernstein, *Short Term Trading*.

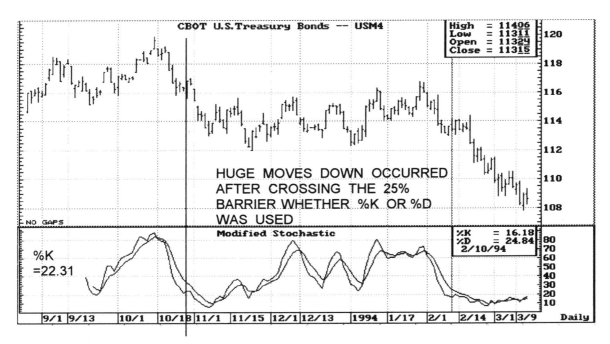

CHART 7-1

THE MACD:

Other traders use the MACD (Moving Average Convergence Divergence) to indicate extremes in market Movements, or worse yet, as a divergence tool. As you know from CHAPTER 5, this is a cleverly designed and highly capable Trend-indicating Oscillator, not an Overbought/Oversold tool. There is, however, an innovative technique (Bernstein, *Short Term Trading*) of using the distance between the slow and fast lines of the MACD as an Overbought and Oversold indicator. In my opinion, however, there is a much better way to accomplish this.

THE RELATIVE STRENGTH INDEX (RSI):

The RSI is not it. Although far better than the Stochastic or the MACD for OB/OS analysis, this important indicator was created by Welles Wilder (Wilder, *New Concepts*) to have universal appeal and easy application across markets. He certainly achieved those objectives, but for the more sophisticated trader something was lost. Since the RSI is *normalized* at +/- 100, like the Stochastic, it squashes strong market moves. If the Oscillator is at 95 and a big up move continues, it only has 4.9999 points left to go.

When the RSI achieved 96.50 on the daily coffee chart below, where could it go? As the move exploded in price, the RSI went down to 93.78, while the Detrended Oscillator increased from 9.41 to 16.19. A seven period input was used for both indicators in this example. On the right side of the chart, we see the Detrend at a whopping value (in relative terms, over four times the original amount of 9.41). The RSI was actually less than it was originally at 89.00. Discernible values of Overbought and Oversold relative to price are of critical importance. Why not use an indicator that plainly illustrates them?

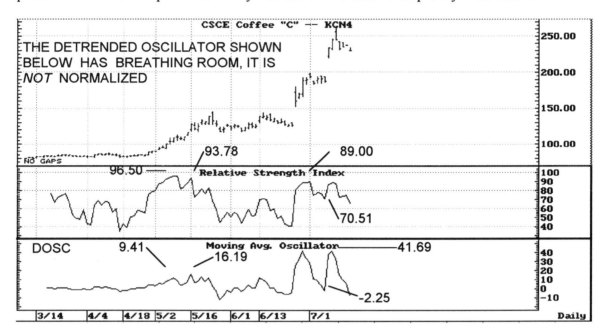

CHART 7-2

Also, consider that Overbought in coffee has a different *character* than Overbought in corn, which has a different *character* than Overbought in the S&P! This character doesn't have the elbow room to show itself using the normalized RSI.

THE COMMODITY CHANNEL INDEX (CCI):

Finally, there's the Commodity Channel Index (CCI). I have the least amount of criticism for the CCI, probably because it does a fairly good job of approximating the Oscillator I use. Although Donald Lambert's[3] development of this indicator was tied to Trend and cycle work, most traders who use this indicator use it as an Overbought/Oversold tool. The CCI is not normalized to +/- 100 and therefore requires more understanding to employ. This is likely why it is not used (or misused) extensively. Although the CCI has value, I believe the Detrended Oscillator achieves significantly better results.

---

[3] Donald Lambert, "Commodity Channel Index: Tool for Trading Cyclic Trends", *Technical Analysis of Stocks & Commodities* magazine, July/August 1983, page 120-122.

THE DETRENDED OSCILLATOR:

The Detrend is an indicator that's been around a long time. I don't know who the originator was or when it was developed. The Detrend attempts to measure variations of price about a zero line which represents the Trend, hence Detrended. We define the Trend as a given Moving Average, then we mathematically make that average constant, or the zero line.

The formula for the Detrend is simple:

Detrended Oscillator = Close minus Moving Average.

A reasonable variation is the high or low, minus a given Moving Average as depicted on Chart 7-5.

Some of my colleagues believe incomprehensible math equates with genius, and therefore to profits. I always believe in keeping things as simple as possible. Since I did my research on the indicator in the early 80s on an 8088 processor, keeping the math simple was a practical, as well as a philosophical consideration.

I approached research on the Detrend in the same way as I approached research on the DMAs. I observed the quality of the indicator's *usefulness, in trading situations*, over a broad spectrum of data. I used none of the typical optimization techniques popularized some years later.

In observing literally thousands of data sets, with a wide variety of combinations of the detrend (simple, weighted, exponential, and mathematical MAs, of the median, high, low, close, etc.), my final conclusion for the best data sets were:

1. The close (today) minus a three day simple Moving Average of the close.
                                   and
2. The close (today) minus the seven day simple Moving Average of the close.

Of the two data sets, the 7 day MA of the close is clearly the most useful in the context I apply it. I still use both data sets, however, particularly under Strategy 1, described below.

Aside from the profits generated from this laborious enterprise, the most gratifying aspect is that I see no reason to change the parameters today, *over 15 years later!*

USING THE DETRENDED OSCILLATOR:

Now, let's talk about how to use this powerful and versatile Oscillator in a variety of easy-to-apply strategies.

STRATEGY 1:

When your position reaches 70, 80, 90, or 100% of average Overbought/Oversold, take your profit.

The key considerations for employment of Strategy 1 are:

The Time Frame we use to calculate Overbought/Oversold and... the definition of what is meant by average Overbought/Oversold.

Here's where experience comes in. *I always calculate OB/OS levels on a daily basis*, i.e. daily data, even though 80% of my trades are off a five minute chart. Let me put this a little differently, so there is no misunderstanding. I never use intraday charts to calculate OB/OS levels for determining Logical Profit Objectives, *even though intraday charts are where my position may be taken*. To determine OB/OS levels, I look back over the Oscillator peaks and valleys. I consider about six months of the most recent daily data.

Average Overbought/Oversold is a value judgment, not a strict mathematical calculation. If I have three Overbought peaks as in Chart 7-3 with values of 96.85, 101.00, and 100.70, I'd take approximately 98.00 as an average Overbought.

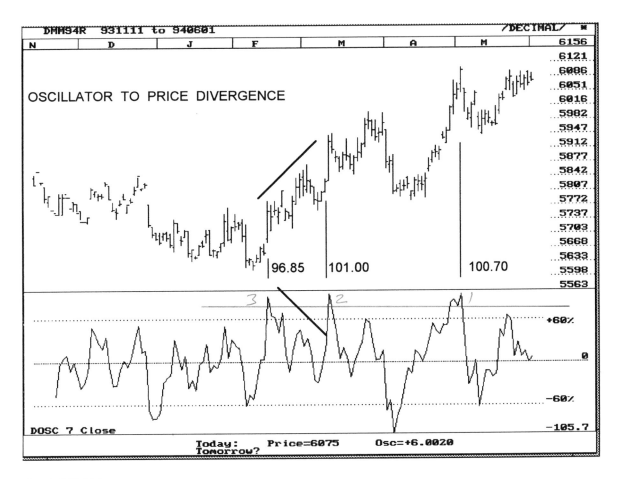

CHART 7-3

I'll typically have my order resting in the market at a *price* equivalent to approximately 90% of the Oscillator's average Overbought level. At that point, I say "adios" to the trade. You might choose a lower or higher percentage. A resting day order is always best to take advantage of unexpected news or large traders pushing the market around for their own purposes. If it doesn't get hit, you cancel it or just let it expire.

Okay, so you have the Oscillator value in mind where you wish to take your profit. You can't call the floor and tell them to take you out at a seven day Detrend of 88 (90% of 98); you need a *price*. To get that price *ahead of time*, you need the Oscillator Predictor™ which I'll describe in some detail at the end of this chapter and in Appendix G. If you don't have the Oscillator Predictor™ to *precalculate* price levels which correspond to the Detrended Oscillator levels you wish to act on, you have another option. Some analysis software (Aspen Graphics™ and TradeStation®, among them) will allow you to set an alert at a given level on an indicator. You hear a beep, you exit the trade. While this is acceptable, the problem with setting an alert on the indicator is that you might miss the trade by the time you hear the alert, act on it, and contact the floor. These price levels are by definition unstable. These prices typically do not last for long, unless the market is in a runaway mode. It is conceivable that if you give a *price order* when the alert goes off, that you will only get out in runaway markets, i.e. those situations that would enhance your profit by *staying in!* So, if you use an indicator alert to get your exit point, simply exit at the market, and hope for the best. Those Mercedes and Jags in the basement parking garage of the Chicago Merc aren't there by accident. Market orders are one of the reasons the locals are able to buy them, so beware.

If profits were taken each time extremes were reached in Overbought, as shown on Chart 7-3, you would not suffer the draw downs of the subsequent pullbacks. These pullbacks are where most traders would have been stopped out by improperly tightening their stops. These pullbacks are for *reentering* against Fibonacci retracement levels. They are not for *exiting!*

When you employ the strategy of utilizing Logical Profit Objectives properly, your percentage of winning trades should increase dramatically, but you may not be there if the market really takes off and keeps going. You could, of course, hold multiple contracts and take Logical Profit Objectives on only a portion of the contracts you hold. It might interest you to know that I have run parallel accounts in which I have taken LPOs on all positions, partial positions, and none at all. Over time, the hands down winner is taking LPOs on *all positions.*

> There's a corollary for Strategy 1. Let's say you're trading an intraday Time Frame and you are using Fibonacci Profit Objectives. The strategy would be to take close-in Objectives (COP's), when operating at or near these extremes in price, as defined by the Detrended Oscillator. A variation of this technique is used even by floor traders I've taught. Their actions in the pit change significantly as these Overbought and Oversold levels are approached. Your actions can change too. Think about it. If a market reaches 70% to 90% of average Overbought on a given day, it is likely that the market will meet resistance and at a minimum, consolidate for the next several days. Under these circumstances, avoid "buy stopping" old highs. Look for, and position yourself, on dips intraday, against Fibonacci support areas. Then, immediately exit against old highs, when the price nears average Overbought again, or when you get close to a Fibonacci Objective, whichever comes first. Remember, the price levels that produce OB/OS will change each day. This is a dynamic condition and dynamically calculated market levels are generally hands down winners over those that are statistically calculated, such as fixed money stops.

Now, I understand that some of you may hesitate to take Logical Profit Objectives because you lack adequate entry techniques that will get you back into the market, once you exit. Part of that problem is addressed below. Most of it will be answered when we study advanced Fibonacci Analysis, DiNapoli Levels™, CHAPTERS 9, 10, and 11.

STRATEGY 2:

The Detrended Oscillator can be used as a filter for *any* entry technique.

I had a high percentage of poor entries before I knew about, and understood, high quality entry techniques. An entry is not only poor if it ends up in losses, it's also poor if it puts you under *significant pressure* before the market goes your way. I dramatically reduced these unfortunate situations by the use of the Detrended Oscillator to determine what value of average Overbought/Oversold was apparent at my prescribed entry levels. If the price level at entry exceeded approximately 65% of Overbought/Oversold, I simply wouldn't take the trade. If the signal stayed in effect the following day, I would again look at the Detrend, to see if the trade was now "safe." If you're unsure of how to set up the Detrend on your equipment, you can try this. Go to your Oscillator Set Up menu, take the one day Moving Average of the close (which is the close), minus the seven day simple Moving Average of the close. That should work... then, just look at the value of the Oscillator at the time your entry signal is given and see how Overbought or Oversold the market is, *before* you act.[4]

---

[4] If you have the Oscillator Predictor™, these levels can be calculated ahead of time and are automatically printed out in the support resistance table, a feature of the CIS TRADING PACKAGE, TIMESAVER function.

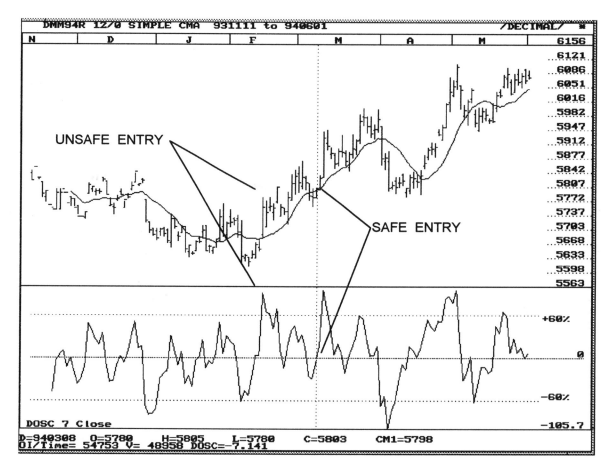

CHART 7-4

Consider Chart 7-4 and imagine a simple system where you would be long when price is above the MA on close and short when the price is below the MA on close. I have pictured a non-displaced 12 day simple MA because that's typical of what many traders use. Two buy signals have been selected for this example.

The thinking here is simple and obvious. If you take a buy signal in a highly OB situation (unsafe), it is likely to produce more pain than one taken at a reasonable (safe) level of OB/OS. In this case, both signals would have produced profit for you, if you weren't bored or frightened out of the trade, i.e. if you stayed with the system criteria. If, however, the system criteria carried an intraday stop, or if the stop was tightened, the unsafe entry may have produced a significant loss.

VOLATILITY BREAKOUT:

I recognize that volatility breakout players might be upset by this strategy, but I think there is nothing wrong with letting the market calm down a bit before entry. Will you miss some good trades? Of course! Will you miss being stopped out many times? Of course! Will the net effect be positive? I think it will. But, test this strategy on your own and see what you think. Many traders are in this game for excitement, not for profit. Some traders can comfortably tolerate 30% or 40% winning trades. I can't. You need to figure out where you fit in.

STRATEGY 3:

An OB/OS level can be used for stop placement.

If you observe the price corresponding to the value of maximum Overbought or maximum Oversold, you can simply put your stop behind that level by some comfortable margin, a few 32nds in the bonds or perhaps 50 points in the S&P. But, you have to be careful how the order reads. If you are using the close minus Moving Average, as suggested above, you should have a *stop close only*. If you want a physical, intraday stop, use the high or low minus the Moving Average, depending upon whether you are short or long the market. These Oscillator values will be respectively higher and lower than the Oscillator calculated using the close, and will produce correspondingly different stops. (See Chart 7-5). I'll guarantee you one thing, your stop won't be located where other stops are, unless by accident. Also, your stop is dynamic, every day it's moved. It should go without saying, that you're within money management parameters, and that whatever signal placed you in the trade remains in effect.

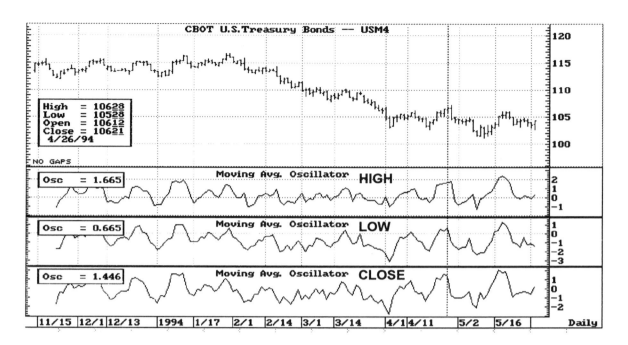

CHART 7-5

The best time to use Strategy 3 is when you have a lot of confidence in your entry techniques. This implies that you don't want to be bothered by close stops and that you choose to give these methods the time and space they require to work. I'll give you two possible examples.

Larry Williams has come up with a variety of high risk, potentially high profit, non-judgmental systems, usually based on some form of pattern recognition. One problem however, is that some of these systems carry a "stop close" only, or no stop at all on the day of entry. This makes some systems users understandably uncomfortable. An alternative would be to hide your (intraday) stop, as suggested in Strategy 3.

When we get to Fibonacci techniques, you will see on occasion, the desirability of an initial, far away, or disaster stop. Strategy 3 can provide an answer to the question, where should the stop be placed? This stop is almost never hit. If the original entry signal is negated, I simply exit "at market," or at the first retracement in the Direction of my entry. Then I cancel the disaster stop.

Realize that if the market is Oversold, a maximum Overbought stop will be miles away. But if the market is nearing Overbought, the stop will be relatively much closer. You can adjust the level of stop placement by using a lower percentage of average Overbought and Oversold, but I would suggest *not going* below 70%.

STRATEGY 4:

CHAPTER 6 discussed the Directional Indicator called Stretch. It uses an Overbought/ Oversold maximum Oscillator level, in combination with a major Fibonacci resistance or support area, to locate a trade entry range. This is Strategy 4. Admittedly, countering an existing Trend can be risky. It is, nonetheless, worthwhile because the combined strength of these two powerful indicators is substantial.

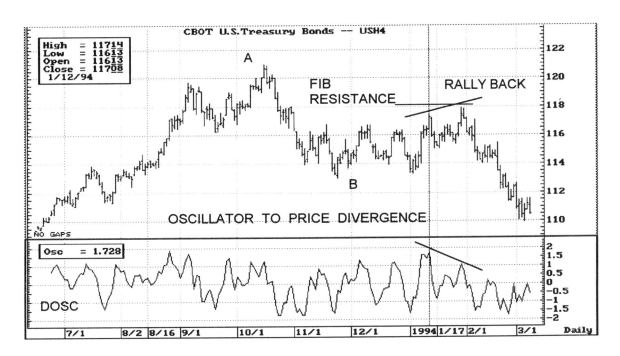

CHART 7-6

Since we haven't covered Fibonacci analysis yet, I will jump ahead a bit, with the following explanation. I suggest that you reread this after completing CHAPTERS 8 through 11, if you don't understand it now. The example above, Chart 7-6, showing daily bonds, was chosen for a couple of reasons. The cursor is placed on the initial Stretch sell. From that point, price reacted down to Fibonacci support and then achieved a Logical Profit Objective (COP) on the upside which was in Agreement with the .618 retracement of the down move from A to B. This is the same Fibonacci Retracement area that helped to give us the initial Stretch sell. To reiterate, *if* at approximately the same price of maximum Overbought, *you have a significant Fibonacci resistance area*, act on this combined level as an entry range to sell.

IMPORTANT POINTS TO NOTE:

The initial Stretch sell gave us two hard down days. *It was the setup signal for the break.* Do you realize how much you can make on a high probability break, where all the intraday trends are in your favor and you can margin up on your selling? If you have high confidence, you can pile up on size and walk away with the loot. You don't need a five point down move to score big; what you do need is patience for the set up and confidence in what is unfolding. In this case, the intraday down trend after the resistance was met, would have supported an aggressive short position, all the way to a support level which was measurable *ahead of time* by techniques covered in CHAPTERS 9 and 10 .

On the rally back to the marginal new high, we only managed to achieve a move back to major Fibonacci resistance, as noted above. The Detrended Oscillator was still high and you then had an opportunity for a second shot at the short side. When we determine ahead of time, the level of Overbought or Oversold, corresponding to a given price level, we can make an informed and *confident decision* about how we wish to handle a given trade.

Every rose has its thorn, and stop placement, utilizing Stretch, can be prickly, since you are countering an existing move. If, for example, your entry sell level was at approximately the close minus MA maximum level, a safe stop may need to be placed at the high minus MA maximum level or behind another, more distant Fibonacci area. I never place money stops in the market. If a safe stop exceeds my money stop, I just don't take the trade.

Since we're on the topic of Oscillators, now is a good time for a prohibition. Don't use an Oscillator to price divergence as an entry technique, unless you have an excellent means of filtering it.. In our last example, there was a divergence signal that worked between the initial high and the rally back high, but this is *not* a high probability signal. Even for the best of Oscillators this can, and often does, spell disaster, particularly for the new participant. Look back at Chart 7-2, as well as other charts contained herein. Price and Oscillator divergences abound, and the market continues to stop out the divergence players. Looking back in time, you're sure to find divergences that work, but when you're going forward in time, *as in real trading*, the accuracy of this technique just isn't there.

STRATEGY 5:

Special applications of the Detrended Oscillator to determine major Trend changes.

There are a variety of such applications. I'll offer one example. The idea behind this strategy is that long term Detrended Oscillator breakouts can be more significant than long term price breakouts. Look at Chart 7-7, monthly gold.

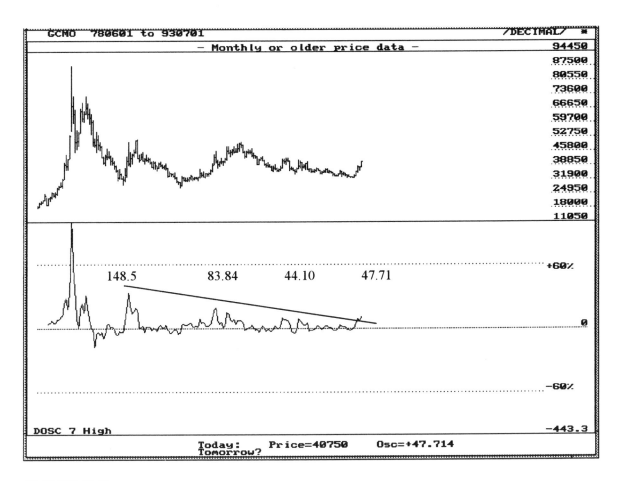

CHART 7-7

For the first time since the 1980 break in the gold market, we have an Oscillator reading, monthly basis, which significantly surpassed previous peaks, corresponding to run ups in price. I am *not* concerned with the divergence the Oscillator has made versus price, but rather with the fact that the Oscillator value has surpassed its previous rally high after a long move down. Note also that when we are measuring momentum in this context, we use the high minus Moving Average or the low minus Moving Average, rather than the close. This is because momentum in this context is attempting to measure the *maximum* push behind the market, rather than its strength at a given point in time.

This indicates that the bear market in gold is over and that we are likely to be in for a period of prolonged consolidation or an up move. To confirm this supposition we would like to see the 47.71 level significantly surpassed in the next gold rally. It would also be nice to see the pullback in gold contained in an ascending pattern. So far this has not occurred.

I've alluded to volatility breakout techniques in CHAPTER 1 and under Strategy 2, in this chapter. The idea behind the success of such methods is that volatility peaks precede price peaks and there is validity in this reasoning. The gold example is really a variation of this. While I will not endeavor to cover fully the details of how I handle extremes in volatility, I will make the following comments so you can understand my general approach. First, you don't know you have an extreme (let's say twice average Overbought or Oversold) until after the fact, so I prefer to take my profits as I near Overbought or Oversold, as stated earlier. When, in hindsight, I see a true breakout in volatility, I employ the techniques in CHAPTERS 9, 10, 11, and 13, to enter *in the Direction of the breakout*. Second, I attempt to filter any such breakout to eliminate blowoffs which will produce amazing volatility breakouts but by definition, the *termination* of any extreme in price.

Finally, as I have suggested, if you are an intraday player, use daily Overbought/Oversold data for calculation of Detrended Oscillator or Oscillator Predictor™ levels. If you trade daily-based, pay attention to daily and weekly Overbought/Oversold levels. If you trade weekly-based, i.e. using a weekly chart, be aware of both weekly and monthly Oscillator levels. Long term mutual fund switching can be greatly enhanced by using these techniques .

Above all, please keep in mind a market axiom I have been stressing for years. It is inherent in my market approach and should be gleaned from the afore-mentioned rules.

LOSS OF OPPORTUNITY IS PREFERABLE TO LOSS OF CAPITAL!

THE OSCILLATOR PREDICTOR™:

In the early 80s, I decided that I needed a means of capturing profit that was more efficient than any I had seen to date. At that time, I was unaware of Fibonacci Expansion Analysis and although the Displaced Moving Averages I had developed gave me reasonable entries, I gave back more "paper profit" than I wanted to in the exit strategies I was using at that time. Any paper profit was to me, *my* profit. I had assumed risk to achieve that profit. I had done considerable work to enter the trade initially and I didn't want the market taking any of it back! The problem simply stated was, "How could I exit on extremes in price, rather than wait for a DMA crossover?" Since I was convinced the Detrended Oscillator was my best Overbought/Oversold tool, and using my background as an engineer, I reasoned that a set of parametric equations could be created which would produce, a day ahead of time, the price level that would correspond to an OB/OS condition in the market. I approached my programmer, George Damusis, with the problem. He crawled off to his office for two weeks and after applying his considerable talents to the challenge, the mathematics behind the Oscillator Predictor™ were created and the study was *graphically* programmed into the CIS TRADING PACKAGE.

Consider what this discovery meant to me. I could accurately forecast a full day ahead of time what level of price would produce a logical (historically-based) profit, a businessman's profit. When you take Logical Profit Objectives, your percentage of winning trades can't help but increase. The main problem in taking logical profits is the lack of knowledge traders have in attempting to re-enter the market at a lower risk point. This issue is addressed in CHAPTERS 8 through 13.

Appendix G shows an example of how the Oscillator Predictor™ works in practice.

SUMMARY

Before we get into the Fibonacci work, let's do a quick summary of the overall plan, (CHAPTER 3), and see what we've learned so far:

THE NECESSARY ASPECTS OF A SUCCESSFUL TRADING APPROACH

1. Money and self management
    See the Bibliography and the Reference material sections.

2. An understanding of market mechanics
    References on market mechanics have been made as we have progressed to this point. When appropriate, more references will be made. For more information, see the reference materials.

3. Trend and Directional analysis
    Lagging and Coincident Indicators theory has been covered.
    We've learned how best to identify Trend.
    We've learned about certain very powerful Directional signals that overrule Trend.

4. Overbought/ Oversold evaluation
    The theory has been covered.
    We've learned how to effectively filter and quantify trades.
    We've learned how to take certain Logical Profit Objectives.

THE BOTTOM LINE IS:
    WE NOW KNOW WHETHER TO BE *LONG, SHORT,* OR *OUT* OF A GIVEN MARKET, AND WE HAVE A MEANS OF DETERMINING WHETHER OR NOT A TRADE WILL HAVE A *REASONABLE* EXPECTATION OF A GAIN.

5. Market entry techniques (Leading Indicators)
    Next we'll see how to position ourselves as safely as possible, within a market that we have chosen to enter according to the above criteria. We will also investigate powerful stop placement techniques.

6. Market exit techniques (Leading Indicators)
    Then, we will cover additional methods of determining Logical Profit Objectives.

SECTION 3

DINAPOLI LEVELS

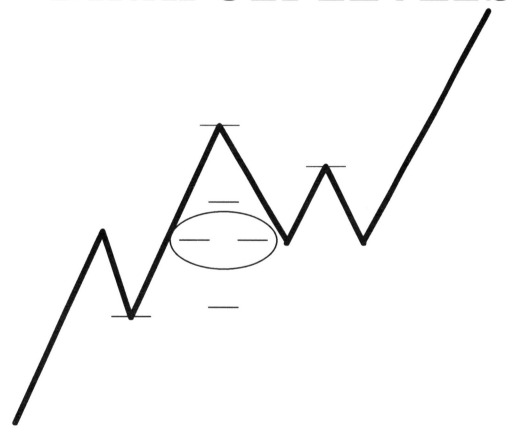

All of today's markets have something in common. Whether we are dealing with futures contracts, foreign exchange, equity markets, or even mutual funds. Speed and volatility have increased *dramatically*. What used to be a days range in the S&P in 1983 is now apparent in one bar on a five minute chart! Regardless of what you are trading, yesterday's methods are ill-equipped to handle today's challenges. The proper use of Leading Indicators address this issue.

CHAPTER 8

FIBONACCI ANALYSIS, BASIC

ooooooooooooooooooooooooooooooooooooooooooooooooooooooooooooo

GENERAL DISCUSSION:

Perhaps in the years to come you might use a portion of your trading profits to travel to some of the great cultural centers of present and past civilizations. If you do, you will find as I did, that Fibonacci relationships are intrinsic to the architecture of Athens, Rome, Amsterdam, Paris, Egypt, many areas of South America, and so on. There's an elemental resonance to the way these aesthetically pleasing shapes derive from these mathematical road maps.

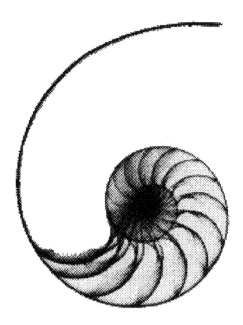

You can observe a curious evolution of Fibonacci expansions played out in musical progressions, crystal formations, and even in the growth of rabbit populations. Whether in the DNA spiral, the preprogrammed construction of a bee hive honeycomb, or the inspirational great pyramid of Giza, Fibonacci relationships abound. The human body itself is a study in Fibonacci relationships. Recently, at one of my workshops I met a surgeon who had done his graduate thesis on facial reconstruction. The thrust of his study was to quantitatively relate post surgical success in appearance with how closely the reconstructed bones approximated the "Golden Mean." There is no denying that Fibonacci ratios and the numbers that generate these ratios are innate in *all* matter.

It takes no great leap of mental agility therefore, to expect that the *combined activity* of mankind will somehow follow these precepts. This is particularly apparent relative to the markets, since markets are linked so closely to the overriding human emotions of *greed and fear*.

A LITTLE HISTORY:

Leonardo was born the son of Guilielmo Bonacci, a wealthy merchant in Pisa about 1170. In Italian, "figlio" means "son," hence "figlio Bonacci" which was shortened through the years to Fibonacci. Signior Leonardo Bonacci was a preeminent mathematician of his time. He's credited with the discovery of the numerical sequence and ratios that have come to be known as the Fibonacci series.

Discovering the Fibonacci series is sort of like discovering America. I'm sure the Indians knew about it before Columbus did. Likewise the proportions defined by the mathematical relationships so important to us as traders, have been around for quite some time.

The Golden Mean or Golden Ratio 1.618, to 1 (or .618 to 1) which among other things closely approximates the cost of this book, has had many names. The Greeks designated the ratio by the letter "phi." Pacioli, a medieval mathematician, named it the "divine proportion." Kelper called it "one of the jewels of geometry." Somewhere along the line, someone dubbed it the "ratio of the whirling squares." I'm glad that one didn't stick. Consider the implications for the title of this book: "The Practical Application of Whirling Squares to Investment Markets."

DERIVATION:

The Fibonacci number series has more interesting aspects to it than most of us can imagine or would wish to. While we might get dizzy considering the possibilities, it's a mathematician's hot fudge sundae, and then some. Consider the series as most of us know it. 1, 1, 2, 3, 5, 8, 13, 21 and so on to infinity. We arrive at the series by simply adding the last two numbers together, beginning with 1,1. The ratios come about from dividing the numbers in various ways. If we divided 13 by 21 for example, we get .619 while 21 divided by 13 = 1.615. If we skip a number and divide 8 by 21 we get .381. Conversely, 21 divided by 8 is 2.625. The higher we go in the number series, before dividing, the closer we come to achieving the exact numerical Fibonacci ratios. We never achieve this, however, since there are an infinite sequence of decimals stringing after it. This is known in mathematics as an irrational number.

One interesting aspect of the summation progression is that it doesn't matter where we start. We can take any two numbers, like 5 and 100. Soon we're dealing with the same series.

5, 100, 105, 205, 310, 515, 825, 1340, 2165

1340 divided by 2165 = .6189

2165 divided by 1340 = 1.616

While it is known that Mr. Fibonacci "discovered" the series after a trip to Egypt, I get a different image when I think of him. Imagine the son of Bonacci sitting under a tree some time in the 13th century, after consuming a great bowl of pasta. He'd likely run out of fingers and toes and had to go to his abacus when the lights suddenly went on. It must have been something like the way I felt when I began applying his discovery to the S&P...as in

WOW!!!

I could go on at great length about the poetry of Fibonacci relationships, but if I did, I'd never get to their practical application to the markets. If you want to pursue the subject, many books have been written discussing the more esoteric aspects[1]. They certainly do a better job of describing the mathematics involved than I have. Besides, my lack of reverence toward the subject is bound to upset some people. So, I will leave the poetry, the derivation, and the history of Fibonacci numbers and ratios to others. This book is about practical application of Fibonacci concepts to the market, and in that vein, I will *bring us back to reality* and repeat the following prohibition for those of you who have galloped ahead to this chapter.

FIBONACCI ANALYSIS SHOULD ONLY BE APPLIED IN THE PROPER *CONTEXT*, WITH PROPER *TRAINING*, AND AS A PART OF AN OVERALL *PLAN.*

---

[1] Many sources of information on Fibonacci-related mathematical concepts are listed in the reference section located at the back of this book.

GUIDE-POSTS:

What *will,* and what *will not,* be covered in the following chapters.

WILL:

$ Basic Fibonacci Expansion and Retracement Analysis applied to the price axis.

$ My Interpretation of Advanced Fibonacci Expansion and Retracement Analysis applied to the price axis, i.e. "DiNapoli Levels™."

WILL NOT:

    Any application whatsoever of Fibonacci analysis to the time axis
    The utilization of Fibonacci numbers in any way (I use certain ratios only)
    Fibonacci Ovals
    Fibonacci Arcs
    Fibonacci Spirals
    Fibonacci-inspired Bands
    Fibonacci-inspired Trend lines
    The *comparatively* minor Fibonacci ratios such as .09, .146, .236, .5.0, 1.382, 2.618, etc.

The topics that I'm not covering are interesting. Some have merit in their own right. However my experience, research, *and direct application to trading* strongly suggest that they aren't worth the time they take to employ, particularly at this point in your learning curve. Since they unduly complicate the picture, I will concentrate only on those ideas and concepts which I have found to be the most useful *and practical.*

BASIC RETRACEMENT ANALYSIS UTILIZING THE TWO MAJOR RATIOS .382 AND .618:

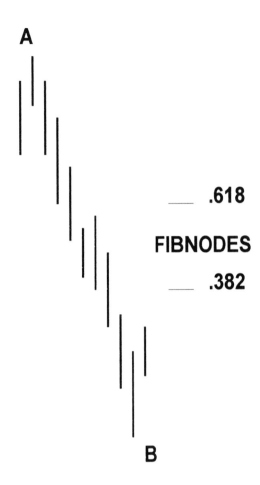

CHART 8-1

Chart 8-1 depicts a down move from point A to point B. Retracement theory states that you measure the vertical distance of the wave between these two extremes of price, (points A and B) and calculate the .382 retracement of this move. At that point, there will definitely, and without doubt, be resistance (selling) to any up move.

Retracement theory does not say that prices must stop there, only that there will be *significant resistance to further movement.*

If price action exceeds the .382 retracement level and continues up, there will definitely, and without doubt, be additional *significant resistance* at a .618 retracement of the same down wave. Will the market stop there? We don't know, but if you were looking to sell that market (context), either point would be an excellent level to initiate sell orders. If you were looking for stop placement areas, hiding your stop behind either level would be a far better approach than picking an arbitrary money stop.

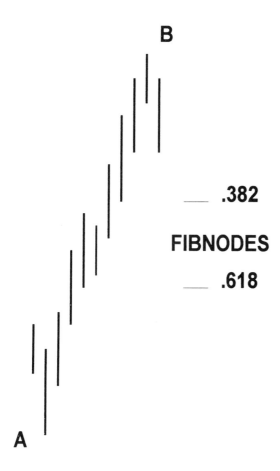

CHART 8-2

In Chart 8-2, we likewise have a move from Point A to Point B. This time it's up. Fibonacci Retracement Analysis tells us there will be predictable support at a .382 and a .618 retracement of the extremes of this wave.

Here are the equations relating to the above criteria.

FIBNODE EQUATIONS

$$F3 = B - .382(B-A)$$

$$F5 = B - .618(B-A)$$

F3 is the 3/8 Fibnode or .382 retracement .

F5 is the 5/8 Fibnode or .618 retracement .

These are slang representations of retracement levels that are traceable back to early Gann work, which referred to 1/8 points as support or resistance levels[2].

Fibnodes are numbers generated from the application of the above equations. Two (or more, as you will see) Fibnode pairs are created per market swing (wave). They will elicit support if they are approached from above or resistance if approached from below.

NOTE: I said *extremes of this wave*, not closes, not hourly medians, not the average of the last two highs before the standard deviation crossover of... get the idea? It's the high and low you are interested in!

---

[2] For those of you who have heard me speak in person or in taped workshops, you know I often refer to Fibnodes as 3/8ths or 5/8ths Fibnodes.

For clarity, idealized bar charts will most often be represented as line charts in the following chapters. These line charts will always be drawn between the extremes of the move, (lows to highs, highs to lows) within a Market Swing. See Chart 8-3.

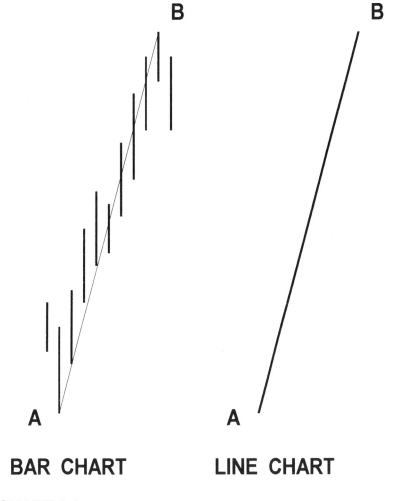

BAR CHART   LINE CHART

CHART 8-3

If you are ever confused about applying the advanced Fibonacci techniques (DiNapoli Levels) taught in this book, create a line chart that accurately mirrors the bar chart you are analyzing. That simple procedure will shorten your learning curve *significantly*.

BASIC FIBONACCI EXPANSION ANALYSIS UTILIZING THE THREE MAJOR EXPANSION RATIOS .618, 1.0, AND 1.618:

The mathematical relationships described by Expansion Analysis, control or specify the growth patterns of price, thereby providing you with Logical Price (profit) Objectives. We refer to these as Objective Points (OPs). Three targets or objectives can be calculated from any ABC Market Swing. The initial thrust may be up or down as in Chart 8-4. Typically, C is within the Market Swing of wave A B but that it is not an absolute necessity.

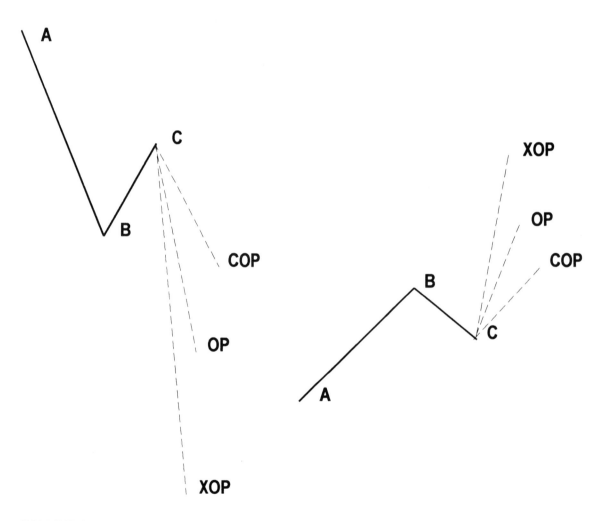

CHART 8-4

The formulas below are used to calculate these profit objectives. The price chart can then be extended from Point C in the direction of wave A B. These extensions are shown as dashed waves signifying the likely progressions of price.

OBJECTIVE POINT EQUATIONS

OP = B - A + C   OBJECTIVE POINT

COP = .618 (B -A) + C   CONTRACTED OBJECTIVE POINT

XOP = 1.618 (B - A) + C   EXPANDED OBJECTIVE POINT

Some previous works on this subject, as well as certain well publicized theory, suggest that the expansion start at the B high or low, rather than at the Point C retracement. My research and experience lead me to disagree with this. Use Point C to begin your expansion. If the expansion on a down wave goes below zero, realize that negative numbers are not "recognized." Otherwise someone would be paying you to take their stock, or their corn, off their hands. Excluding certain tax shelters, I don't know where that occurs.

Expansion analysis says nothing about time. The dashed wave is shown achieving the various objectives at different times, for clarity only. In fact, it is possible for a given AB wave to reach all three Objective Points, after some reaction, likely against Fibnodes. In cases of strong Directional moves, price may go immediately to the XOP. The strength of the market during the AB leg, as well as the lack of strength or depth of the retracements on the BC leg, help us to determine which of the three price objective targets is initially met. OP stands for Objective Point; COP for Contracted Objective Point, since it is the smallest of the three possible objectives; XOP for Expanded Objective Point, since it is the largest. Generally speaking, OP targets are met more often than COP targets, before a significant retracement occurs. XOP targets are least frequently fulfilled.

There are other valid Fibonacci expansion ratios, but a trade off must be made between excessive clutter and proven reliability. My research and experience shows the expansions stated above are most reliable and most worthy of our attention. This should become apparent when you see *how these ratios are used, combined, and applied* in the next chapter DiNapoli Levels.

FREQUENTLY ASKED QUESTIONS:

Could you discuss the concepts of Fibonacci arcs?

Although my area of expertise is Fibonacci analysis, that doesn't mean I *use* all aspects of applying this analysis to the markets. My specialization involves the practical application of Fibonacci analysis to the markets. In other words, *how can we make money out of this.*

I researched Fibonacci arcs back in 1989. They were not useful enough for me to include in my trading methodology. If you wish to pursue this path, the reference section contains material on this and related subjects.

You use a 7 day oscillator, a 7x5 and a 25x5 Displaced Moving Average but 7 and 25 are not Fibonacci numbers. Why do you use them?

I don't care if they're not Fibonacci numbers! Seven and 25 work in the studies you cited. I'm not a Fibonacci zealot. *I use what works!*

Why does Fibonacci theory work?

To some extent, it is a self-fulfilling prophecy since certain knowledgeable entities, both large and small, successfully employ it. That is not a sufficient explanation however. Fibonacci theory is natural law. All of us have our own tolerance for risk, pain, and fear. We also feel degrees of greed. While each of these feelings are expressed in varying degrees, the average of these emotions for a crowd are somehow quantified by these mathematical relationships, and faithfully expressed in the markets.

CHAPTER 9

DiNapoli Levels™

□□□

INTRODUCTION & CAUTIONS:

DiNapoli Levels are applicable, with uncanny accuracy, from one minute charts to yearly charts, perhaps longer. If you plan to trade the very short term, be prepared to work hard, very hard. Even though computers, used in conjunction with the proper software, can ease this workload, diligent attention wears all of us down. Regardless of the quality of the approach or the thoroughness of your analysis, it's easy to burn out, and in the process squander hard won equity.

Throughout this book, I am attempting to include not only the product of my research, but also the knowledge and experience I have gained from hard won trading lessons. I feel it's a part of my mandate to wave red flags at the sand traps. Since I've been teaching a long time, I've had ample opportunity to observe both the successes and failures of my students. These observations have led me to the following cautions. Once you have digested the material and have learned to apply DiNapoli Levels properly, you need to be watchful of overconfidence and the consequential lack of money management that results from a series of *unbelievable calls*. Aside from that, sloppiness in the application of context, as well as lack of experience with order entry and floor operations, are additional *serious pitfalls*. If you are buying and selling in the wrong place, it doesn't matter if your floor knowledge is limited. You'll lose anyway. When you buy and sell in the *right place* however, there is great competition for fills and you *will* experience problems you may have never encountered before[1].

---

[1] Joe DiNapoli, "X'd Trade or Where's My Fill?" *Technical Analysis of Stocks & Commodities,* March 1995, page 88.

ELLIOTT WAVE PRINCIPLE:

Many believe that Elliott Wave Principle is synonymous with Fibonacci Analysis. It is not. Fibonacci analysis can stand on its own (not recommended) or as part of an overall trading strategy which includes or excludes Elliott Wave. Those of you who are familiar with the Elliott Wave Principle will quickly find particular value in the trading approach taught in this book. Knowing where you are in a particular Elliott Wave count is confusing at best to long time practitioners, let alone newcomers. DiNapoli Levels circumvent the problem and look at waves as expansions and contractions of themselves, period.

DINAPOLI LEVELS™:

In CHAPTER 8, I described *basic* Fibonacci analysis. Some form of this basic approach is what most people practice when using Fibonacci analysis on the price axis. That's fine as far as it goes, but to progress to a deeper level, we must have a variety of definitions to clarify and quantify our thinking. Understanding these definitions is an absolutely essential prerequisite to your command of the subject matter and for your eventual application of advanced Fibonacci analysis (DiNapoli Levels) to the markets. So, refer back to these definitions as many times as necessary, until you can call them your own.

DEFINITIONS:

MARKET SWING:

A Market Swing is a *trader-defined* market move, lasting minutes or years, taken from a "significant" market low or high, which occurred sometime in the past, to the most recent high or low. A Market Swing can be referred to as a wave. In the following chart the Market Swing would be between the Focus Number and Reaction 5.

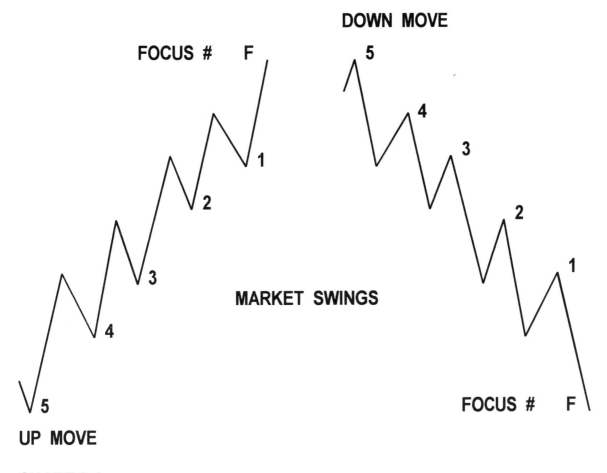

CHART 9-1

REACTION NUMBER or POINT:

A Reaction Number usually is a low or high point within a given Market Swing (numbers 1 through 5). I have avoided using the terms "swing high," or "low," in the definition of Reaction Numbers for two reasons. First, some of you will attach certain inapplicable qualifiers to this term.   Second, we will have Reaction Numbers in areas that "swing highs" and "lows" cannot possibly be found.

There can be multiple Reaction Numbers within a Market Swing.   What determines whether Reaction Numbers are lows or highs is the Movement of the Market Swing.   In our example, each swing has five Reaction Numbers.   For our purposes, Point 5 is considered to fall within *our* definition of Reaction Numbers, even though Point 5 might not be *reacting* from any previous point.   It could be, for example, an all time low price, as with hot, Initial Public Offering (IPO).   In fact, Point 5 has very significant importance, as it is the extreme of the Market Swing.   It is referred to as Primary Reaction Number or indicated by the '*' symbol .

FOCUS NUMBER:

The Focus Number is the extreme of the Market Swing.   It is the location on a chart, from which *all* retracement values (Fibnodes) for a given Market Swing, are calculated.   If the Focus Number changes, *all* Fibnodes for a given Market Swing change as well.

FIBNODE or NODE:

A Fibnode or Node is a number based upon Fibonacci retracement ratios, which will elicit support as the market approaches it from above or resistance as the market approaches it from below.   Two Fibnodes, or Nodes, are calculated, one at a .382 retracement, and the other at a .618 retracement between the Focus Number and a Reaction Number.

FibNodes™ is also the name of a software program used to calculate and present Fibonacci retracements and objectives.   *Within this text*, except for headings, FibNodes will appear with a capital F and a capital N when referring to the software program.

OBJECTIVE POINT:

An Objective Point is a number based upon Fibonacci expansion ratios, which marks a targeted Profit Objective for an advancing or declining wave.

CONFLUENCE:

Confluence ('K') is a price point or area which occurs when two Fibnodes from *different* Reaction Numbers have the same, or almost the same numerical value.   The Confluence must occur only between .382 and .618 Fibnodes.   An *area* of Confluence would include the Fibnodes that create the Confluence, as well as the range of price between them.

Confluence presents significantly stronger support or resistance than a single Fibnode. Confluence (closeness) is dependent on the volatility and Time Frame of a Market Swing. Fibnode Confluence can therefore be widely disparate from one chart to the next. For example, the extremes of the price range of a one minute or a monthly chart are incredibly different. Likewise the price range in a given Time Frame may vary widely. We might have a range of price in one day of 250 points, and 1250 points in another day. Confluence is subjective. It keeps the programmers and non-judgmental traders confused, and that's healthy for the longevity and usefulness of this approach.

LINEAGE MARKINGS:

Lineage Markings[2] are semi-circular arcs used to visually identify which Reaction Numbers create a given Fibnode.

LOGICAL PROFIT OBJECTIVE:

A Logical Profit Objective is a predetermined price point where orders will be placed in the opposite direction from that which you are trading. If you're long, this will manifest as resistance. If you are short, this will manifest as support. Two Logical Profit Objective location techniques are Oscillator Predictor Points, and Fibonacci-derived points. Fibonacci-derived points can be those that come from Fibonacci expansion analysis, or, as you will see, levels created from certain Fibnodes singularly, or at Confluence levels.

Since trading is simply a game of percentages, it should follow that accurate (Logical) Profit Objective Points would significantly increase your ability to evaluate your percentage, your chance of continued profit!

AGREEMENT:

Agreement is an area of price which occurs when the proximity of a Fibnode and an Objective point (COP, OP, or XOP) is "acceptably close."

---

[2] In some workshops and in the TRADING COURSE, these markings have sometimes been referred to as cross hatch markings.

FIB SERIES:

A Fib Series is the *combined set of Fibnodes*, created from the proper application of DiNapoli Levels to the price axis. It is *not* the Fibonacci Summation Series discussed under Basic Fibonacci Analysis in CHAPTER 8.

DINAPOLI LEVELS™ or D-LEVELS™:

DiNapoli Levels are support and resistance levels created from a specific set of rules, governing the advanced applications of Fibonacci analysis to the price axis. DiNapoli Levels include Fibnodes, Objective Points, Confluence, and Agreement price areas.

EXAMPLES:

The first step in determining DiNapoli Levels is the proper location of the Focus and Reaction Numbers.

Let's take a look at the following chart:

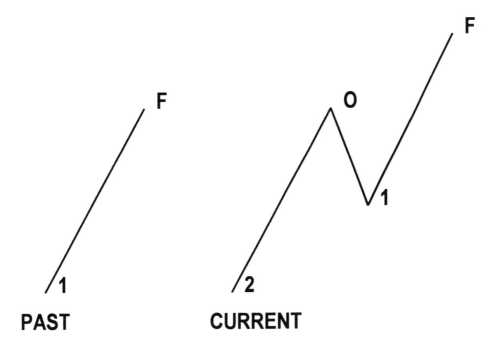

CHART 9-2

In Chart 9-2, we're looking at the progression of a Market Swing over time. The Focus Number is the high of the move. In the PAST wave, Reaction 1 is the Primary Reaction Number. This wave would produce two Fibnodes.

The old Focus Number at O in the CURRENT wave, has no significance in determining the *new or current* Fibnodes. The CURRENT wave has two Reaction Numbers, 1 & 2. Reaction Number 2 is now the Primary Reaction Number. As we learned in CHAPTER 8, there are two Fibnodes per Reaction Number, therefore we can now generate four Fibnodes.

The following Chart 9-3 shows four Fibnodes, a Confluence area, and Lineage Markings. There is Confluence 'K' in the area created by the .618 reaction of the F to 1 leg, and the .382 reaction of the F to 2 leg.

It is crucial for you to know which Reaction low created each Fibnode pair. They must be clearly associated. Fibnode Lineage tells us a lot about the nature of our response when that Node is approached by market action. A bunch of lines splayed across a chart with no identifying Lineage only confuses our action. It doesn't help us. There's no need to place Lineage Markings up to the Focus Number, since all Fibnodes in a Market Swing are created from the *same* Focus Number. That's why it's called the Focus Number.

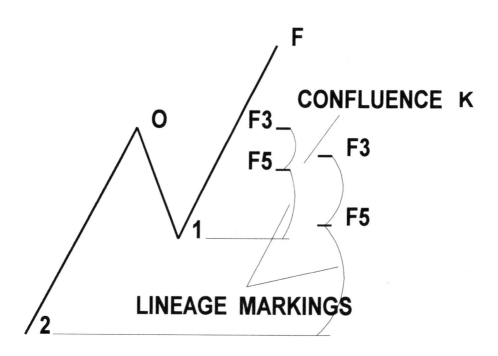

CHART 9-3

Chart 9-4 is an example of how Agreement arises:

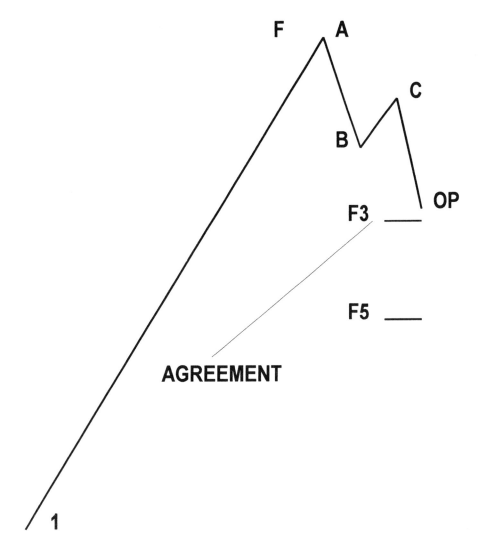

CHART 9-4

A .382 support Fibnode between the Focus Number 'F' and Reaction 1 is acceptably close to an OP expansion from C. Acceptably close is subjective. It is dependent upon Time Frame and volatility.

Now let's look at a slightly more complex wave.

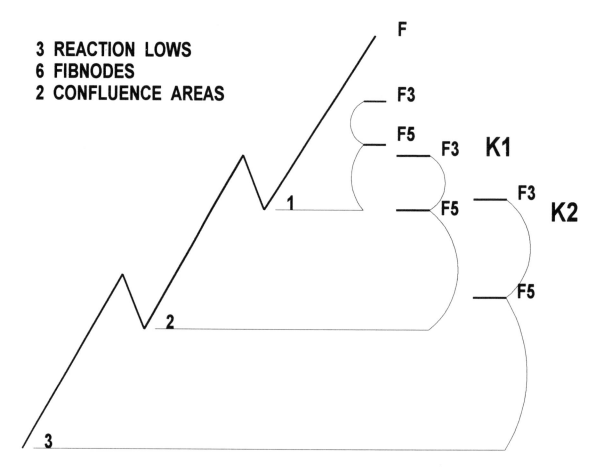

3 REACTION LOWS
6 FIBNODES
2 CONFLUENCE AREAS

CHART 9-5

Chart 9-5 is obviously an up wave. A down wave would be annotated in a similar manner, consistent with the charts shown earlier.

In this up wave, we have two Confluence areas, K1 and K2. If you choose to be a buyer as the market progresses upward (context), it would be an excellent strategy to buy just above the Confluence area K1, and hide your stop just below the K2 area. The specifics of order entry will be covered later in the Fib tactics section, CHAPTER 13.

Now let's look at a real market example, so you can see how effectively the concept of Confluence works in real life trading. Below is Chart 9-6, the monthly deutsche mark.

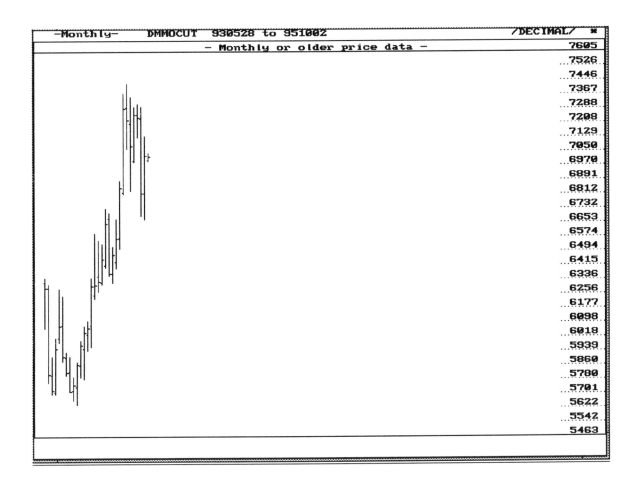

CHART 9-6

If we label this as shown earlier in the idealized Chart 9-2, we'll have four Fibnodes, two of which create the Confluence area shown in Chart 9-7.

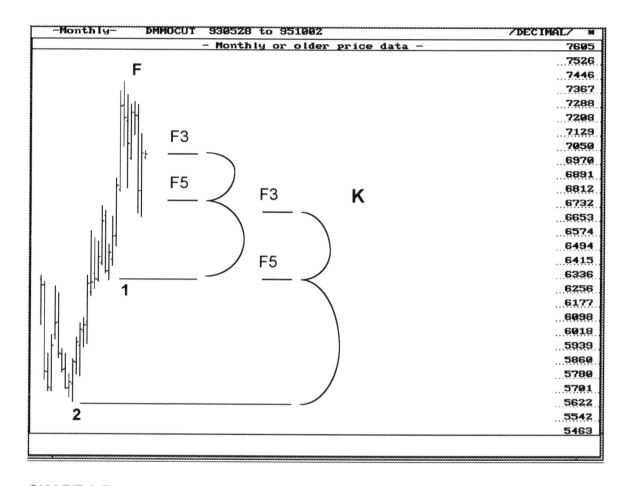

CHART 9-7

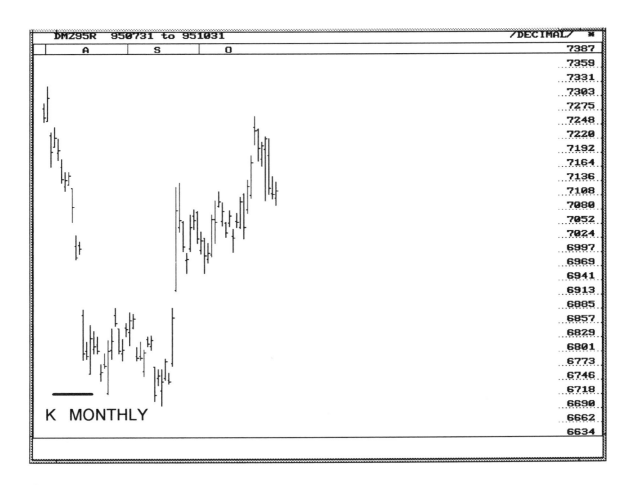

CHART 9-8

As expected, the Confluence area provides substantial support and sets up the market for a major rally. See the daily Chart 9-8. If we didn't know about the monthly Confluence area and we were short this market, we would be in for a surprise. From our definition of Logical Profit Objectives, this area of Confluence would have been a great place to close our shorts. If we were monthly-based players, we could then wait for a retracement back to reinitiate our short positions (as long as our context for the trade remained in effect). If we were daily-based players, there are a number of ways we could get long the daily move. I'll consider a few.

    1. The rally off of Confluence looks suspiciously like a daily Double Repenetration. If it were (not shown) we'd jump on the long side like a duck on a June bug. We could do this in a number of ways.

        A. Buy the first shallow retracement, after the Confirmed Double RePo.

       B. Buy stop the high, after the smallest of pull backs, after the Confirmed Double RePo.

       C. Anticipate the Double RePo, and enter the market after a pull back, but with some Confirmed intraday Trend in our favor.

  2. Enter the market on a retracement, after a Confirmed Trend up in the Time Frame of our choice.

  3. Look for a Directional signal to support our trade. Enter on the long side accordingly.

> While it was not specifically mentioned in our Directional Indicator chapter, (you wouldn't have had the basis for it), a strong Confluence area, particularly on a weekly or a monthly chart *could* be a Directional signal on its own. This approach is a bit risky, unless the market is oversold as in "Stretch." The way I employ this strategy is to wait for a Confirmed Trend, in this case up, then employ one of the Fib tactic entry signals we will cover in CHAPTER 13. A number of my students will "Bonsai" into a Confluence area. I typically use either Minesweeper A or B. My suggestion to you, is *not* to jump ahead now to see what these tactics are. I think it's better for you to reread this section later and stay with the chronology of the book.

IMPORTANT POINTS TO NOTE:

1. If you didn't precalculate these DiNapoli Levels on the monthly chart, you wouldn't have any idea that support was about to manifest. The market was falling like a stone. You should *always* study the higher Time Frame charts, so you know where you are in the bigger picture.

2. If you look at the daily (not shown) in the vicinity of the first (highest) Fibnode shown on Chart 9-7, you would have seen a nice playable rally.

3. On Chart 9-7, if you re-label 2 to A, 1 to C, and the high preceding 1 to B, F would have been a near perfect OP.

4. There will be a much more compelling example of Confluence coming up later in CHAPTER 11 involving the 500 point single day Dow decline. It's delayed until later because it is a little more complicated. It involves more reaction lows and consequently more Fibnodes. I realize it's best to learn to walk before we start to run.

THE PROPORTIONAL DIVIDER:

Let's back up a little and talk about how to get DiNapoli Level markings placed properly on a price chart without spending much money.

You can use a precision architectural tool called a proportional divider (precision ratio compass):

The proportional divider[3] is the least expensive way to *properly* pinpoint DiNapoli levels, since it allows you to identify Fibnode locations and Lineage, quickly and easily. I do *not* recommend the use of graphics (technical analysis) software[4] to accomplish this task, since you'll get a series of unidentifiable lines splayed across a chart. This type of presentation will degrade your ability to confidently act on the information at hand.

If you want to use a computer to accomplish this, you have a couple of choices. With enough talent, understanding of the concepts, and *foresight of your trading needs in the heat of battle*, you may be able to adequately program a spread sheet. Otherwise our FibNodes™ software produces a tabular printout with identifying characters. This presentation accurately characterizes the type of Fibnodes you are encountering in current market action. The characters act similarly to Lineage Markings. Using the software does not dispense with the utility of the divider. A high-quality proportional divider is great to have around. If you are trading intraday however, *it is effectively impossible* to accurately keep up with things, without adequate software.

BOTTOM LINE: Use a divider and/or adequate software for best results!

---

[3] Coast Investment Software, Inc. offers a high quality, wide span, lightweight proportional divider, along with the Applications Manual which shows you how to properly use the divider. Dividers can sometimes be located in architectural supply stores. There is a wide variety of these devices. Be careful which one you purchase, not all are appropriate for use in this context!.
[4] This prohibition goes for our own CIS graphics package which has a good Fibonacci graphics study. It too however is inadequate for the trader to efficiently and accurately implement DiNapoli Levels.

CHAPTER 10

DiNapoli Levels™
MULTIPLE FOCUS NUMBERS & MARKET SWINGS
▯▯

Now that you have a firm foundation in my approach, it's time to progress to the next level of difficulty. We'll begin this more advanced level of analysis with a Market Swing that is very typical of what you will see occurring in the market place every day. See if you can properly label the Focus Number(s) and Reaction Numbers on Chart 10-1, before turning the page.

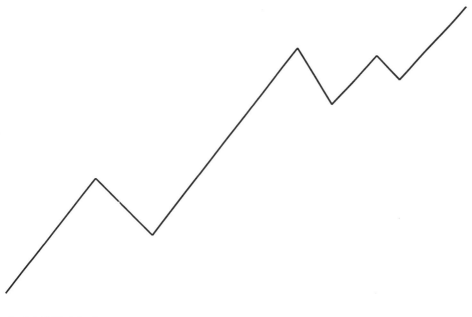

CHART 10-1

With the exception of points M and Q, does your labeling look like this?

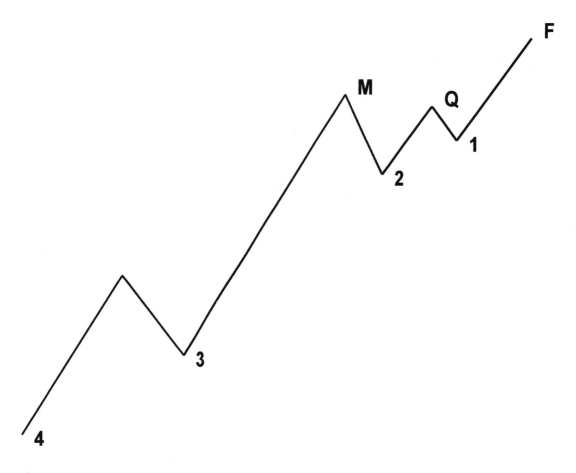

CHART 10-2

Here's what the DiNapoli Levels would look like.

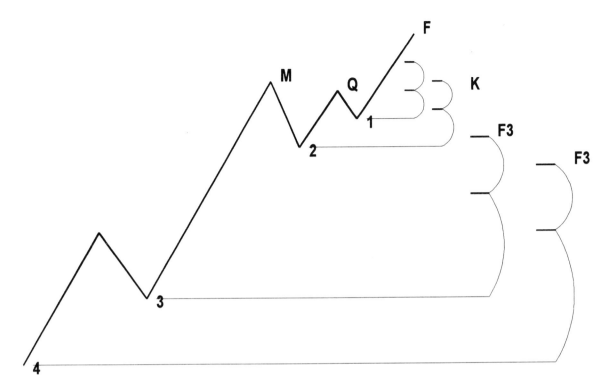

CHART 10-3

Even though point Q did not take out the high at M, the reaction low labeled '1', is still significant and should be used to create DiNapoli Levels.

Notice there was a Confluence area 'K' formed between the .618 Node of Reaction 1 and the .382 Node of Reaction 2. There could be no Confluence between the .382 Nodes formed from Reactions 3 & 4, since they were both .382 Fibnodes. If the .382 Node off Reaction 4 and the .618 Node off Reaction 3 were a bit closer, then we'd have another area of Confluence 'K'.

MULTIPLE MARKET SWINGS:

Now let's look at a chart that has two Market Swings, one up, one down.

Can you properly label the Focus and Reaction Numbers on this chart?

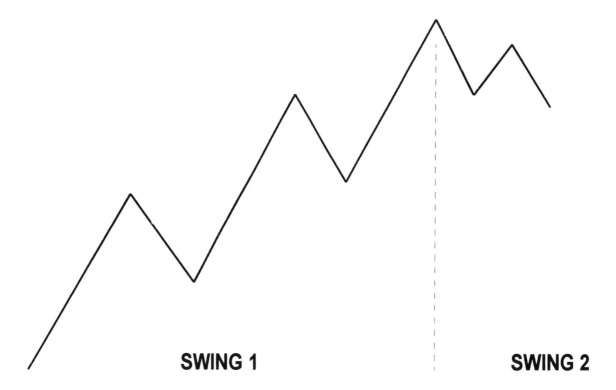

CHART 10-4

In Chart 10-5, we have two Focus Numbers, Focus support Fs, and Focus resistance FR. Focus support numbers produce Fibnodes that elicit support, while Focus resistance numbers produce Fibnodes that elicit resistance.

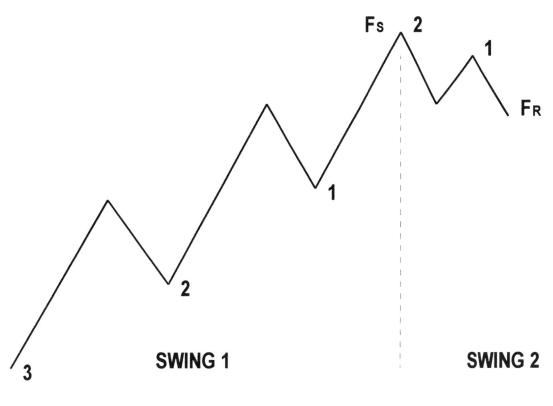

CHART 10-5

The following Charts 10-6 A & B show a portion of Chart 10-5.

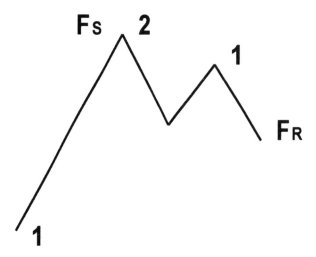

CHART 10-6A

I'm going to assume that you can properly produce the D-Levels for the up swing. This would include a total of six (support) Fibnodes. For the sake of clarity, in Chart 10-6B, I show only two of the support Fibnodes, i.e. those associated with the portion of the up swing from Point 1 to Fs. All of the resistance Fibnodes for the down swing are shown.

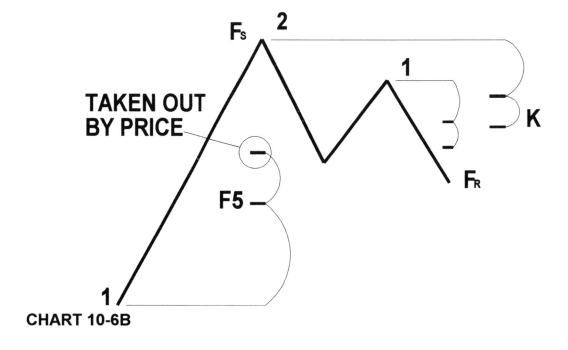

CHART 10-6B

The first (highest) support Node survived the *first* corrective move down from Fs, even though the Node was briefly penetrated. It was clearly taken out by the subsequent move down, point FR on the chart. This first support Node is no longer active and can be removed from our consideration. It has been *taken out by price.* Cleaning up your charts, as well as your software tables (spread sheets or FibNodes) is essential for maximum clarity and optimum decision making.

The down move from Fs to FR has two reaction highs, four Fibnodes and one Confluence 'K' area. Excluding the possibility of some Directional Indicator leaning on price Movement, it is likely the price will bounce around between the support Node at F5 and the resistance Confluence area 'K.' Knowing that this action is likely on a short Time Frame chart allows for the possibility of scalping.

ADVANCED COMMENTS:

Let's go back and talk about this brief penetration of the .382 support Node on the *first* corrective move down from Fs. If it were 1986 or 1987, I would have considered that this Node was a dead issue, no longer in play. Now however, with more individuals using even crude Fibonacci analysis, there are sometimes large numbers of stops placed just under *certain* Fibnodes. These stops act like gasoline strewn around the pit and knowing floor traders as I do, they aren't shy about lighting the match. Since this Node was only penetrated briefly and then supported, I consider it to still be *active.* *If we had the capability*, we would drop to a lower Time Frame to see just how sharp the penetration was. The briefer the better, to consider the Node as active. We're not finished yet. We could also check out the possibility of there being an OP move down from the high at Fs to the vicinity of the first support Node. *If* this were the case and the Agreement range between the .382 support Node and the OP move down, takes us to the low formed before the move back up to 1, there is certainly nothing wrong with the Node. It is still in play! All we have done is fulfilled the OP. A third consideration when we drop the Time Frame, would be to look at reaction lows buried in the up move from 1 to Fs. It's possible there is a reaction low in the vicinity of the first support Node, presenting the floor with a group of tasty sell stops to go after.

MORE FOCUS NUMBERS:

We just had an example of two active Focus Numbers existing on the same chart, one eliciting support Nodes and one eliciting resistance Nodes.

What about three Focus Numbers - is that possible?

Try labeling Chart 10-7 on your own.

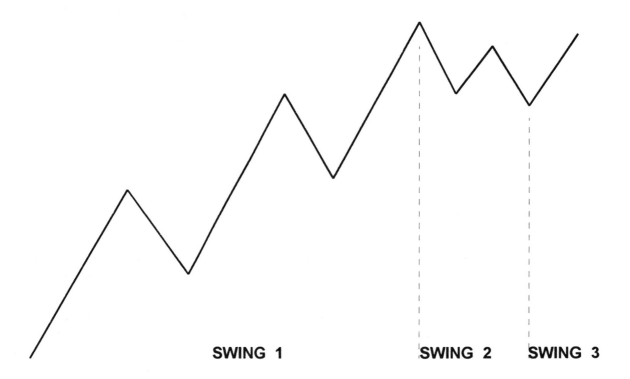

SWING 1   **SWING 2**   **SWING 3**

CHART 10-7

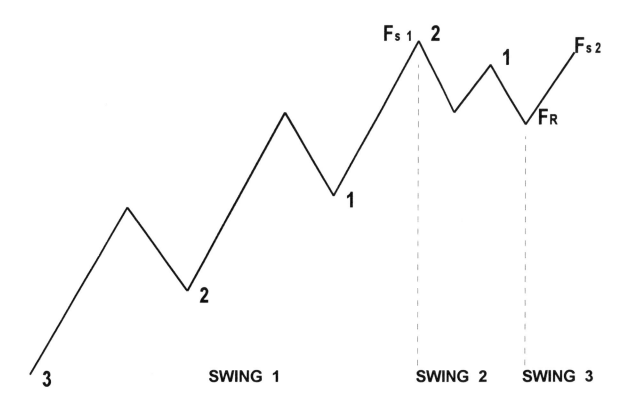

CHART 10-8

If you are having problems labeling these charts, note a couple of rules that will help you out.

1. You will have as many Focus Numbers on a chart as you have Market Swings.

2. Reaction Numbers are always "behind," or earlier in time, than the Focus Numbers that they are associated with.

3. For a Focus number to "own," or to be associated with a given reaction low in an *up move*, the Focus Number must be the highest high after the reaction is manifest. That's why Fs2 only has one reaction low, not four! Fs2 is lower in price than Fs1.

For a Focus Number to "own," or be associated with a given reaction high in a *down move,* the Focus Number must be the lowest low after the reaction is manifest.

If Fs2 broke to a price level below FR, we would be back to two Focus Numbers, Fs and FR, as shown in Chart 10-9.

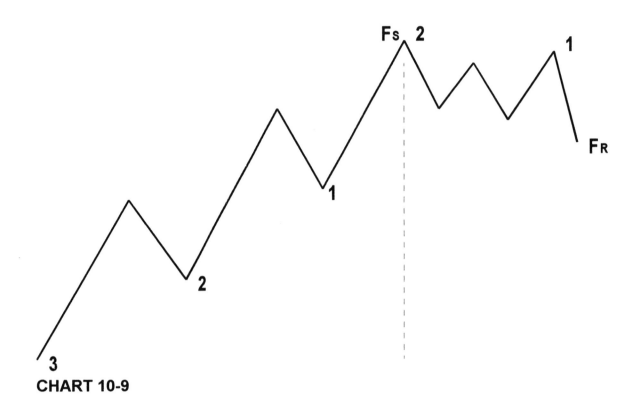

CHART 10-9

TIME FRAME'S IMPACT ON FOCUS NUMBER:

If we then lengthened the Time Frame significantly, as in Chart 10-10, the resulting Market Swings would show up as much simpler waves. Think about it; if you go from a five minute to an hourly chart, there will be fewer discernible retracements, i.e. *Reaction Points will disappear.* It's the same situation any time you increase your Time Frame. Likewise by reducing your Time Frame, you will get more Reaction Points, more Confluence areas, and more opportunity to fine tune your entry and profit objectives.

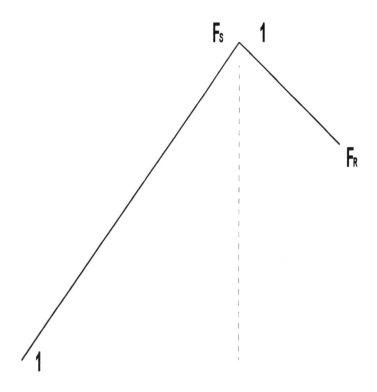

CHART 10-10

We'll discuss more about moving the Time Frame around and what it will do for you, in the next chapter. If your eyes are crossing, don't worry about it, there will be lots of examples coming along. Once you discipline your mind to follow the preceding rules, these ideas should fall into place with amazing ease. If you are wondering how I came up with this theory, it was from trading a five minute S&P. When support came in where I didn't expect it, I'd work my way back and calculate the likely Focus and Reaction numbers necessary to produce such support or resistance. I then went through a very lengthy verification process. After three years or so I had it down.

LESS IS MORE

If you haven't yet figured out why that statement is true, you're probably still in your twenties, or maybe your thirties. With regard to trading, this is a critical issue. Individuals often challenge themselves up to the limit of their ability and beyond. Some traders believe that keeping track of 23 indicators and 94 support levels provides them with more comprehensive analysis than two indicators and six support levels. In trading for profit, look at more markets if you must, but don't foul yourself up with a spaghetti chart.

PRUNING THE FIB SERIES:

With respect to keeping things as simple as possible, let's look again at how we can *eliminate* Fibnodes as they become irrelevant and thereby simplify and *clarify the path to our objective.*

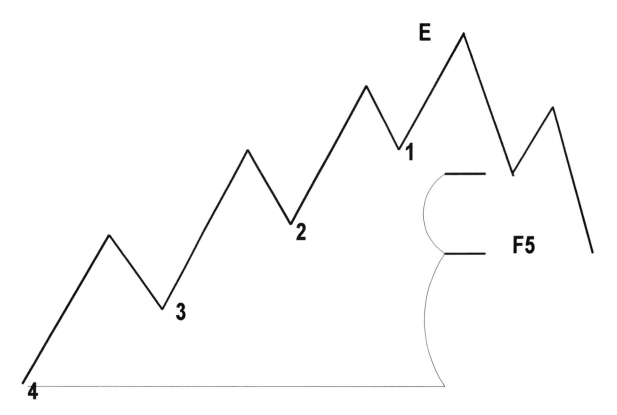

CHART 10-11

Chart 10-11 shows a clear progression of an up Movement with shallow retracements until price reaches point E (Enough). The price then makes an OP move down into Agreement with the .618 Node of the entire up move. That simplifies our life as traders, since we are now able to remove from consideration all of the Fibnodes that the down price action has penetrated through (less is more.) F5, off the primary reaction low '4', is the only active *support* Node left on the chart.

The next Chart 10-12 is a favorite of mine from private tutoring sessions. It's a five minute line chart depiction of the S&P on a day the Federal Reserve cut the discount rate by two points, the First Lady was rumored to have had an affair with the Chief of Staff, and the President was wounded in a related incident. China also invaded Taiwan, Congress approved the capital gains tax reduction and ... well you all get the idea.

CHART 10-12

Try to label this, before turning the page.

There are only two active Fibnodes remaining on this chart which are probably more than the number of traders still solvent. All the others have been taken out by the price action.

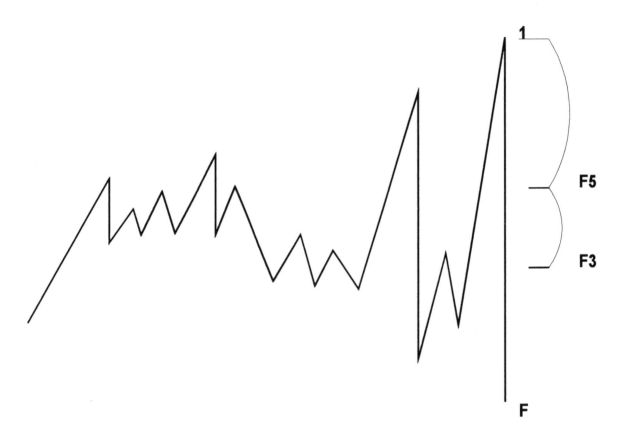

CHART 10-12A

CHAPTER 11

TRADING WITH
DiNapoli Levels™

GENERAL DISCUSSION:

Your success in trading with DiNapoli Levels is dependent upon a comprehensive understanding of what has been taught throughout this book. Your ability to recognize how different Time Frame charts of the same data impact your trading strategy is of the utmost importance. Spend as much time as you require internalizing this subject. The dividends paid will be many.

MOVING THE TIME FRAME

To some of you this will be absurdly obvious. Others among you will really struggle with the visualization of the way Time Frame interacts with DiNapoli Level creation. If you have trouble with this process, it will become clear as you work with graphics software and display the *same data in different Time Frame charts*.

We'll start very simply. The half hour Chart 11-1A, has an eight price bar up move. The line Chart, 11-1B, shows all the information from the bar chart that we need for D-Level creation. Chart 11-1C shows the proper labeling of Focus and Reaction Numbers for this Market Swing.

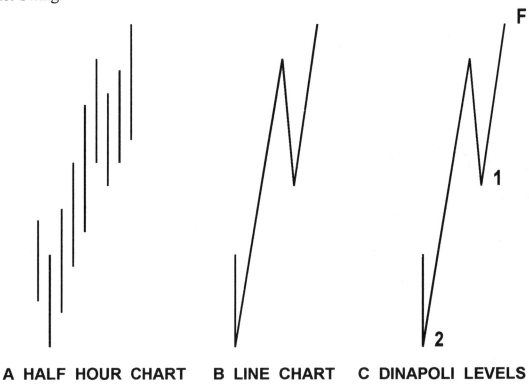

A HALF HOUR CHART   B LINE CHART   C DINAPOLI LEVELS

CHART 11-1

The *same* (up) price Movement depicted on an hourly-based chart has only four price bars instead of eight, since it takes twice as much time to create each bar. In this process a reaction (number) disappears.

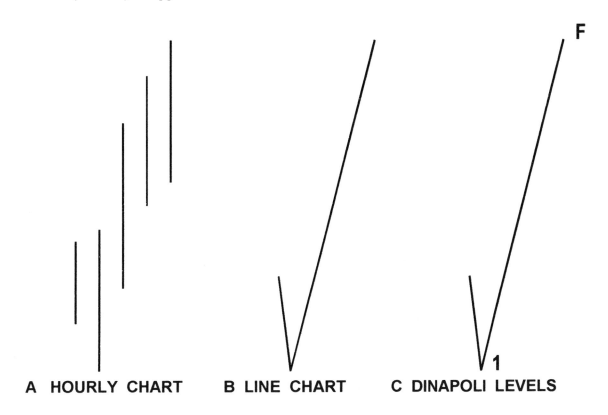

A HOURLY CHART   B LINE CHART   C DINAPOLI LEVELS

CHART 11-2

Both charts are equally accurate for the Time Frame shown. The point is that as you increase the Time Frame:

> five minute to hourly
> daily to weekly
> weekly to monthly, etc.

the number of DiNapoli Levels will likely *decrease*, because the number of Reaction Points are likely to decrease.

Conversely if you decrease your Time Frame:

> monthly to weekly
> weekly to daily
> hourly to five minute, etc.

you will likely increase the number of DiNapoli Levels.

Let's say you are a daily-based player, but your equipment gives you the ability to collect hourly data (as is the case with an inexpensive delayed feed). You can produce more Fibnodes on an hourly basis than you would otherwise see on the daily chart. You can therefore fine tune entry areas and stop placement points. You can still do your analysis at the end of the day, but with an hourly capability you can create additional Fibnodes from Reaction Numbers that would be buried in the daily action. These additional Fibnodes may create areas of Confluence transparent to higher Time Frame traders.

AN IDEALIZED TRADING EXAMPLE:

Let's try to apply some of the techniques we've learned so far, in an everyday trading situation. We'll assume we've dropped the Time Frame sufficiently to locate a Confluence area, as shown on Chart 11-3. We'll also assume the following criteria regarding Trend. The Stochastic (not shown) has given a "sell" while the MACD (also not shown) remains in the "buy" mode. The Trend therefore remains intact to the up side. Furthermore, for the sake of simplicity, we'll say the Trend will remain intact to the up side, even if price action were to break to the area of Confluence.

The questions: Where would you enter, and where is your stop?

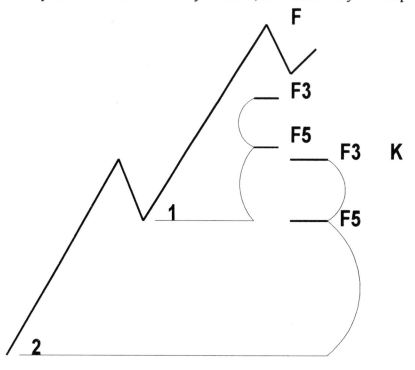

CHART 11-3

Ponder the questions before considering the solutions outlined below. There is more going on here than is obvious. Incidentally, there is more than one correct answer.

Hyper Hank: *Joe, I'd sell the market right where it is and take profits at the Confluence area. Then I'd go long.*

This answer is correct only if Hank had the ability to lower his Time Frame sufficiently to observe the Trend on the Market Swing down. Hyper Hank needs a down Trend to justify his sell, and while a lower Time Frame *may* give an MACD/Stochastic "sell," it's an assumption on Hank's part and beyond the specified criteria. Also, any offset to his short position or any order that would initiate a long position, should not be placed at 'K' Confluence, but rather *above* Confluence in order to increase his chances of being filled. Hank also did not answer the question fully. He said nothing regarding stop placement. He's so anxious to trade, he doesn't consider protecting himself. My advice would be for Hyper Hank to settle down and regroup[1], otherwise he is about to learn an expensive lesson.

Conservative Carl: *Joe, I'd put my buy at the .618 retracement of Reaction 2, and place a stop beyond the old low at Reaction 2.*

This solution assumes the Trend would still be *up* at the .618 of Reaction 2, and the criteria given, only guaranteed an up Trend to Confluence. If we assume the Trend would remain up at his entry point, I would recommend that Carl:

A. Should put his sell stop *at*, not below #2 (assuming he has a broker that commands respect in the pit).

B. Should buy *above* the Primary Node, (which is, as he stated, the .618 retracement of Reaction 2), *not on it*.

If Conservative Carl had qualified his entry, depending on the Trend remaining intact, this solution would be acceptable, but perhaps overly cautious. The problem with waiting for deep retracements to manifest, is that the context (in this case Trend) may be history at the entry point, and Minesweeper A or B would have to be employed for a proper entry. See Fibonacci Tactics CHAPTER 13.

---

[1] He should reread CHAPTERS 4 & 5 on Trend analysis. There are also a variety of trading psychology sources listed in the bibliography, reference, and appendix sections, which would be of value to him.

Diligent Dan: *Joe, I'd enter just above Confluence and depending on my money management criteria, my stop would be below Confluence or below F5 of the Primary Node. If I choose the later stop placement criteria, I'd keep my eye on the Trend. If the Trend broke to the down side, I'd exit at the market or calculate the resistance DiNapoli Levels at that point, and take my first opportunity to exit my long, at or below a resistance Node.*

Good answer Dan, but you are missing something.

Hyper Hank is back!: *I'd put my buy above the first .382 support Node and my stop below Confluence.*

That would be my choice as well, but give me a reason.

I don't want to miss the move!

Diligent Dan is back!: *Hyper's second solution is Joe's preference because there's Agreement between the OP of the down move, and the first .382 retracement area.*

CORRECT! See Chart 11-4.

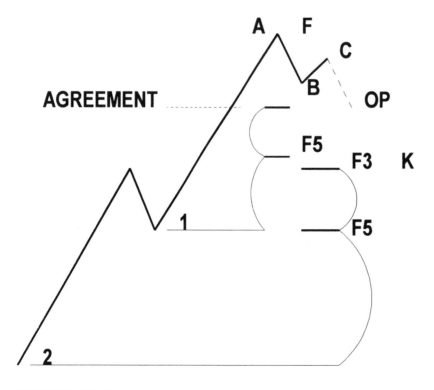

CHART 11-4

You can see from the progression of ideas that there is more than one acceptable answer possible from analysis employing the same methodology.

ADVANCED COMMENTS:

It's probably a little advanced at this point but I'll be more specific about *my* stop placement.

If the higher Time Frame Trends supported a long entry, (better, safer context than what was given), my initial stop would likely be below the primary .618 '*' Node.

If the higher Time Frame Trends did not support the trade, then my stop would be just below Confluence.

If the trade criteria included a Directional Signal up rather than just a Trend up, I would have entered just above the first .382 Node, even if there were no Agreement, since a Directional Signal is stronger than just a Trend up. I would also buy stop the old high, or the high at C *if* the Directional Signal were particularly strong (like a Double RePo Failure). If I were filled on both the 'Or Better' buy (at the first .382) as well as the 'buy stop,' that would be fine. I don't mind doubling size on Directional moves.

D-LEVEL™ EXPANSION ANALYSIS & LPOs:

Now let's look at a somewhat more complex set of Market Swings with respect to Fibonacci Expansion Analysis and Logical Profit Objectives , as in Chart 11-5 below.

See if you can locate *all* profit Objectives on Chart 11-5 before turning the page. As usual, there is more here than meets the eye.

CHART 11-5

HYPER HANK'S SOLUTION:

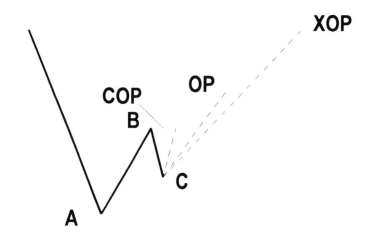

CHART 11-5A

DILIGENT DAN'S SOLUTION:

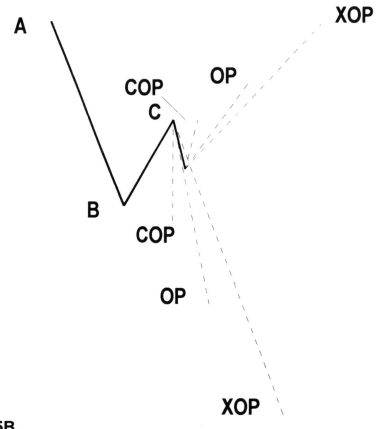

CHART 11-5B

Dan correctly perceived that I did not specify whether we were long, short, or out. In this type of configuration, there are Objective Points both up and down, depending upon the labeling of the A B C move.

Hank's labeling of Chart 11-5 was correct but incomplete. I have not included these labels on Dan's solution for clarity, but I did include all expansions. If I were to include both labels on this one chart, one set would have been labeled A, B, C (Hank), the next set would have been labeled A′, B′, C′ (Dan). Both resistance Objective Points and support Objective Points were created from the combined labeling.

As Diligent as Dan was, he was *not* diligent enough. Do you know what's missing? Think about our definition of Logical Profit Objectives before turning the page.

RESISTANCE NODES AS
LOGICAL PROFIT OBJECTIVES

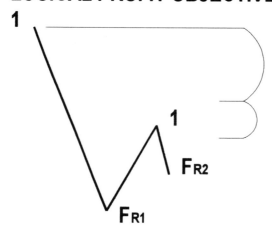

CHART 11-5C

SUPPORT NODES AS
LOGICAL PROFIT OBJECTIVES

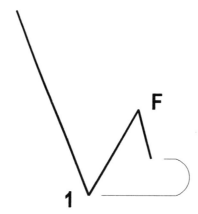

CHART 11-5D

Based on our definition of Logical Profit Objectives, the Fibnode retracement areas should be included, unless these Fibnode retracement areas have been taken out by price. Including these Fibnodes gives you five Logical Profit Objective Points, not three, for a given Market Swing.

In Chart 11-5C, there are two independent Fibonacci Resistance series i.e., two Focus Numbers FR1 and FR2, each "owning" one Reaction Point. For clarity, only the Fibnodes for the FR1 series are shown. If you think the FR2 series is too small to bother with, consider that I have not labeled the 11-5 series of charts with a Time Frame. I have made them small on purpose. If we are looking at a five minute chart, the Nodes created from the FR2 to 1 wave might be safely ignored, but what if this represented a weekly Time Frame?

Chart 11-5D shows a support Node that will act as a Logical Profit Objective for a position on the short side. The .382 Node has been taken out by price and therefore is not shown. The position of the .618 Node is such that the price is on a LPO right now! Which of the LPOs we take, will depend on the *context of the trade* and the criteria we established for our original entry[2].

Here are some context issues to consider when we determine which profit objective to take.

1. How overbought or oversold are we?

2. Are we in this trade because of a Directional Indicator or a Trend Indicator?

3. Are the higher Time Frame Trends with or against us?

4. Are we about to get a confirmed or unconfirmed Trend signal in a higher time frame that will support our trade?

5. For intraday traders, we want to know how close we are to the end of the day!

6. Was thrust apparent on the A B leg?

7. Are we nearing our stress point? Newer traders are often more comfortable with losses than profits. If you feel undue stress because of the level of profitability, take a close in objective before you act irrationally.

---

[2] Now would be a good time for you to reread the Directional Indicator material in CHAPTER 6 relating to "Bread & Butter."

MORE ON STOP PLACEMENT:

We'll take a look at an up move, to consider more thoroughly stop placement techniques. The same reasoning applies equally to down moves.

DiNapoli Levels can have a significant effect on how you place stops, beyond what has already been discussed. Let's take a look at stop placement behind, against, or in the vicinity of old market highs or lows.

Have you ever considered why, on one occasion the market blows through an old high like gangbusters, and on another, the market holds firmly in the *vicinity of,* but not necessarily *at* the old high? Take a look at the following examples.

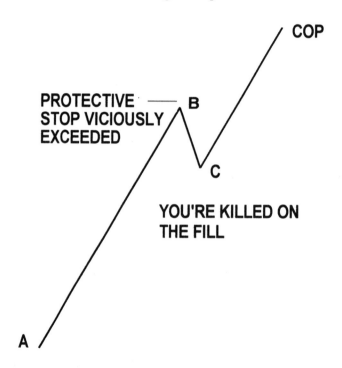

CHART 11-6

Due to the small extent of the pull back of the B-C leg, the COP would have been significantly higher than Point B. It therefore would not have presented any resistance in the vicinity of the previous high at B. If you had a stop in the vicinity of B, and it is exceeded, you would have stood a good chance of being hit hard, since there was nothing to stop the market from brisk Movement up.

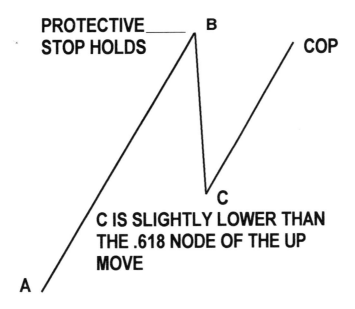

CHART 11-7

In the situation depicted in Chart 11-7, the protective stop is less likely to be hit, and if it is, your fill is more likely to be reasonable. You not only have some inherent resistance from an old market high, but you also have the additional resistance applied from the COP. This helps the high at B to remain intact.

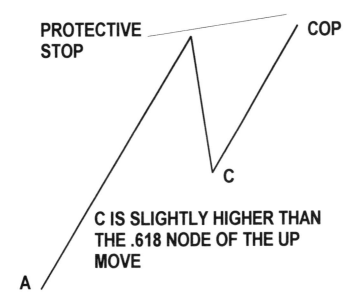

CHART 11-8

In the situation depicted in Chart 11-8, Point C is slightly higher than a .618 retracement of the same AB up move. Stop placement should account for the location of the COP and be placed *above the expansion*, not just above the preceding high.

In Chart 11-9 below, we see a variation of the same stop placement criteria shown in the previous charts. In this case, the up leg is forming a COP, just before the previous high, thereby increasing the chances the previous high will hold.

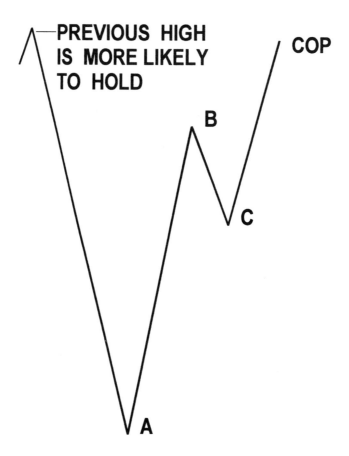

CHART 11-9

This phenomenon is played out in the weekly U.S. Bond Chart 11-10, as the twin tower, all-time highs were formed. COP resistance halted a move to new high ground.

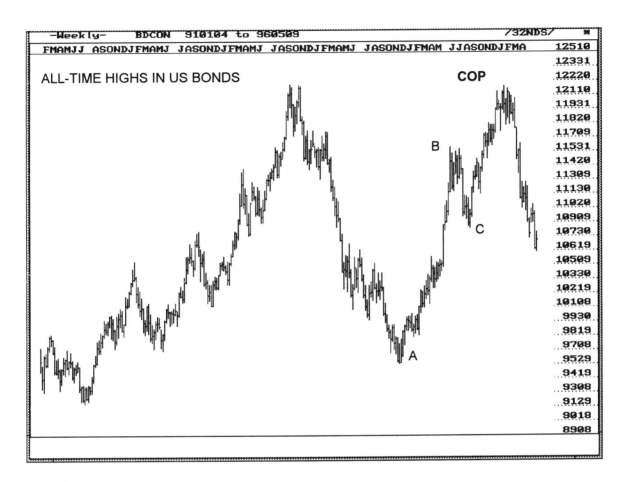

CHART 11-10

Once you become familiar with DiNapoli Levels, you will see a poetry evolve in market moves that is truly fascinating. It's like holding a prism up to white light and seeing the colors of the entire spectrum emerge.

PRESENTATION:

At this point in the book, I think it's fitting to digress a minute and discuss the presentation of the upcoming market examples. This book has been self-published, because I wanted to control the material. If it is accepted as a useful and competent work, I'll be very pleased. If it is not, it won't be because an editor who knows nothing about trading cut the heart out of it. While I have maintained control, such control involves risks.

Presumably, you're studying this book because you want to find out more about my trading techniques, or about the specific topics the book deals with. Up to this point, I've mentioned certain products Coast Investment Software (CIS) offers, primarily as footnotes. That is as it should be.

The safe way to handle the upcoming examples would be to keep the presentation generic. The problem with that approach is that it's not the best method to help you learn, nor is it the best method for me to teach. My work is most efficiently and lucidly taught, if I use the tools I have developed, so that you can see *what I'm doing* The FibNodes™ program is such a tool. The Proportional Divider previously mentioned is another such tool. These are both products that CIS offers. You can get around the use of these tools if you choose to, but my job of teaching the material and your job of learning the material is made much easier by their inclusion in this book.

The FibNodes program allows the user to represent in a tabular format, the Retracement and Expansion points we've been talking about. It also allows for identifying characters to be associated with the Nodes it creates so Lineage can be established. This feature bears repeating for two reasons. First, Lineage is a part of the methodology that some of my students tend to ignore - to their regret. Second, if you choose to use a spreadsheet to implement the concept, you need to build a Lineage feature into it. The FibNodes program has other features designed for high intensity data management during the trading experience and has been designed as a high quality trading tool. The next few pages will explain the FibNodes program printouts *only to the extent that is necessary for you to most easily understand the way I have developed and use Fibonacci analysis.* A list of the FibNodes program features can be found in Appendix F.

FIBNODES™ PRINTOUTS:

For your understanding of the concepts taught in this book, we are utilizing FibNodes software, DOS version 4.32, from which all printouts are generated. The program can handle up to 30 Reaction Numbers. Typical printouts will contain three or four Reaction Numbers. In practice, I seldom use more than 12 per file, since 12 fit conveniently on the monitor and are reasonable to keep track of. Upon entering any Reaction Point, you can enter an identifying character (such as '*') after the last digit of the Reaction Number. This character will be carried through to the Fibnodes associated with the particular Reaction Point you select.

Charts 11-11, 11-12, and 11-13 illustrate the presentation of data in the FibNodes program.

17 Apr 97   13:47:44   Updated: 04/17/1997   .0382 0.618
Focus Number File C-930   Focus# (High for the swing)
Point Number Support Fib Nodes   Point# (Enter highest reaction low first)

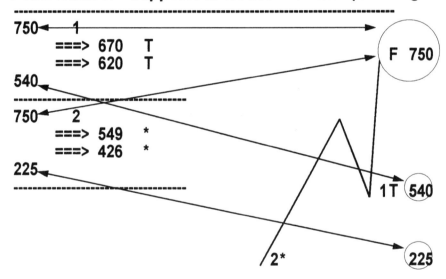

```
750◄──1──────────────────────────────►
      ===> 670   T                           F  750
      ===> 620   T
540
750◄──2
      ===> 549   *
      ===> 426   *
225                                          1T (540)

                                    2*       (225)
```

CHART 11-11

17 Apr 97   13:47:44   Updated: 04/17/1997   .0382 0.618
Focus Number File C-930   Focus# (High for the swing)
Point Number Support Fib Nodes   Point# (Enter highest reaction low first)

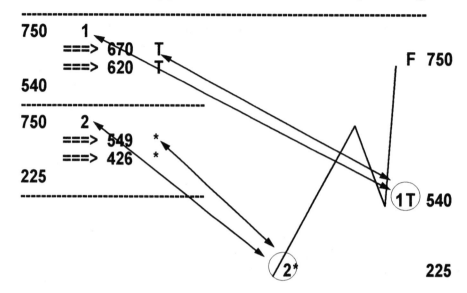

750   1
         ===> 670   T
         ===> 620   T
540
         ----
750   2
         ===> 549   *
         ===> 426   *
225

F 750

1 T 540

2*

225

CHART 11-12

17 Apr 97   13:47:44   Updated: 04/17/1997   .0382 0.618
Focus Number File C-930   Focus# (High for the swing)
Point Number Support Fib Nodes   Point# (Enter highest reaction low first)

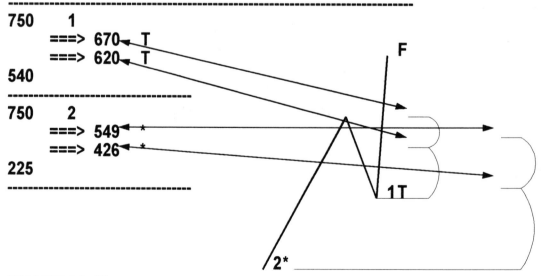

750   1
         ===> 670   T
         ===> 620   T
540
         ----
750   2
         ===> 549   *
         ===> 426   *
225

F

1T

2*

CHART 11-13

We have an idealized example of a Market Swing, starting at 225, achieving a high, then retracing to 540 and thrusting up to 750. User inputs to the FibNodes™ program are 750, 540T, and 225*. They are shown on the left side of the FibNodes printout. The Focus number 750 is automatically entered for each segment (1 & 2) since Fibnodes within a series are always created from the same Focus Number to each Reaction Number. Box 1 contains the .382 and .618 retracements between the Focus Number and the first Reaction Point. Box 2 contains the .382 and .618 retracements between the Focus Number and the second Reaction Point and so on. Whether the FibNodes file is a support or a resistance file, .382 Nodes will always be shown at the top of the box, while .618 Nodes will be shown at the bottom of the box. This is done for two reasons. First, when you are trading, you want to see the first number which is likely to provide support or resistance in the current market situation. Second, the *form* of the printout with the top numbers always .382 Nodes and the bottom Nodes .618, allow you to easily and quickly pick out Confluence by comparing top and bottom numbers. If two reactions are close to one another, they will produce *Nodes* that are numerically close but both Nodes will be on *top* or both will be on the *bottom*. Therefore they are *not* areas of Confluence. I always try to make things as fool proof as possible for my own trading, since stress tends to eat up IQ faster than multi-tasking eats up computer memory.

You can also enter any variety of identifying characters you choose, to indicate the Lineage of a given Reaction Point. 'D' can be used for a *daily* Reaction Point, 'M' for a *major* point. Some Reaction Numbers are more significant than others, so the capacity to enter letters for clarification after the Reaction Point, is of substantial analytical value. In our idealized example, Reaction 2 is followed by '*' indicating that it is the *primary* Reaction Number and 'T' follows Reaction 1 indicating *thrust*. These characters are carried through automatically to the associated Fibnodes by the software.

Finally, naming FibNodes™ files is up to the user but the naming convention I use can help you to figure out what any example is illustrating . Odd numbered FibNodes file names are resistance files and even numbered names are support files. You can get other information as well. In our next example, the Dow file is named DJYR02. DJ is the instrument, YR the Time Frame, 02 indicates support. If DA was in the name, instead of YR, it would indicate a Daily file. A five minute FibNodes resistance file of the September S&P would be named SPU051.

SP - S&P 500
U - September
05 - 5 minute
1 - Resistance

DOW EXAMPLE:

In the Dow example, utilizing the high in 1987 at 2736[3], I placed an asterisk after the 41 Reaction Point, upon entry into the software, since 41 is the primary reaction low, i.e. the depression low occurring, of course, after the 1929 market crash. The low in 1957 was relatively minor, so the small letter 'm' has been inserted to let me know how strong these Fibnodes are likely to be. The low of 777 in 1982 deserves a capital '*M*' since it was a major point, the beginning of this incredible bull market. You can see the sharp ascent after the 1080 low up to 2736. Therefore 1080 gets a 'T' for thrust. Thrust Reaction Numbers are extremely important, for a variety of reasons. The FibNodes printout below details Fibnode support and shows a clear area of Confluence between the Thrust Fibnode at 1712 and the Primary F3 Fibnode at 1707. Those of you who were around and trading during the nearly 1000 points in four day *crash* and 500 point single day decline, know just how gut-wrenching this experience was. We had a negative premium of thousands of points in the S&P cash to futures spread, with no end in sight and suddenly all this was stopped dead in its tracks at 1706.90, by *precalculated Fibnode Confluence support!*

THIS WAS NOT AN ACCIDENT!

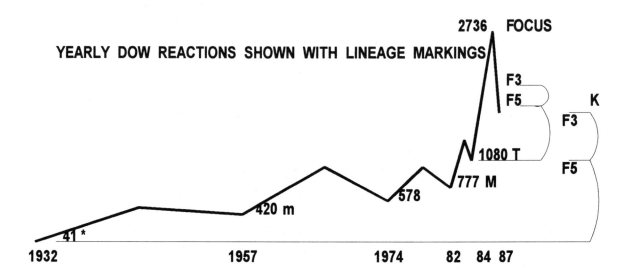

CHART 11-14

---

[3] The Dow values used in this example are the actual *print value,* not the averaged (mythical) high and low theoretical values, as published in some financial newspapers.

22 Apr 97   15:27:24   Updated: 04/22/1997   0.382 0.618
Focus Number   File DJYR02   Focus# (High for the swing)
Point Number   Support Fib Nodes   Point# (Enter highest reaction low first)

2736   1
     ===> 2103   T
     ===> 1713   T
1080

2736   2
     ===> 1988   M
     ===> 1525   M
777

2736   3
     ===> 1912
     ===> 1402
578

2736   4
     ===> 1851   m
     ===> 1305   m
420

2736   5
     ===> 1707   *
     ===> 1070   *
41

Copyright (c) 1996 CIS, Inc.

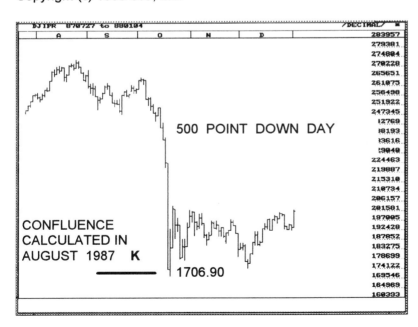

CHART 11-15

FIBNODE™ OBJECTIVE PRINTOUTS:

In addition to providing Retracement Numbers (Nodes) at which to enter trades, FibNodes provides you with Logical Profit Objectives which we refer to as Objective Points (OPs). The three targets, or objectives we've been discussing have printouts that look like this:

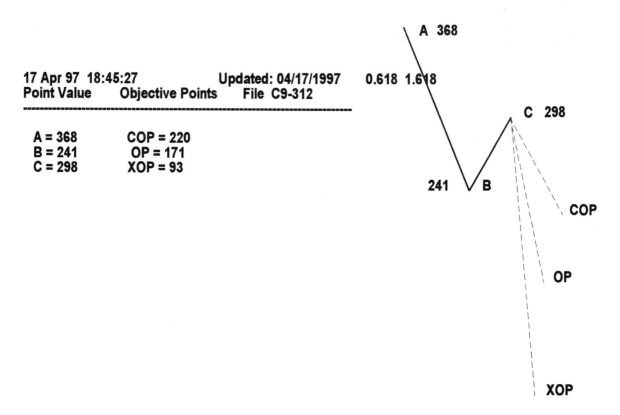

17 Apr 97 18:45:27   **Updated: 04/17/1997**
Point Value   Objective Points   File C9-312

   A = 368   COP = 220
   B = 241   OP = 171
   C = 298   XOP = 93

CHART 11-16

The numbers on the left side of this printout are the values of the A, B, & C points. The numbers on the right side of the printout are the calculated Objective Points.

If you see any extensions on the end of FibNode file names (.FIB .OP), don't be confused. These extensions help the program and therefore the trader to more quickly and easily locate previously generated files.

AGREEMENT ON THE BOND CONTINUATION:

Now that you're familiar with FibNodes printouts, let's consider an example of Agreement that nailed a major weekly low in U.S. bonds. We'll look at the same chart of the U.S. bond market we looked at earlier, but we'll label it in order to determine where likely support will come in, after the double top at the 122 level.

The area of Agreement between the COP (10528) of the A, B, C and the '*' support Node of the up move 1 to F, shown on Chart 11-17, held firmly. Subsequently this area led to a rally in the bonds, from 10528 to a high of 117 a few months later, which was an approximate .618 retracement of the preceding down move! See Chart 11-18.

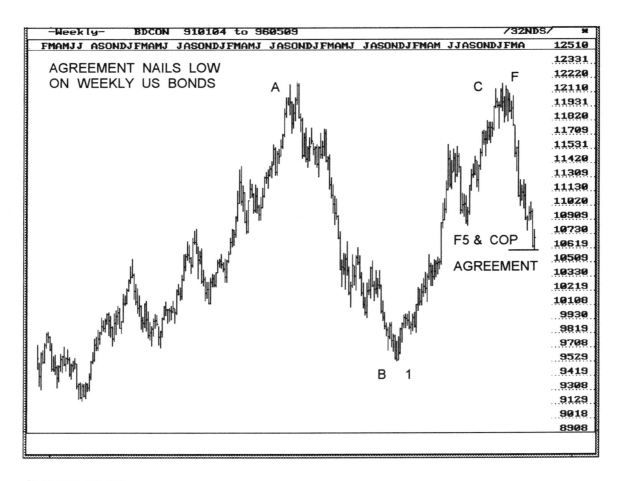

CHART 11-17

These are FibNodes Printouts detailing the support area in Chart 11-17.

```
22 Apr 97  23:37:20              Updated: 04/22/1997      0.618  1.618
Point Value      Objective Points     File  BDWK02       /in 32nds/
-------------------------------------------------------------------

     A = 12210      COP = 10528
     B =  9601       OP =  9527
     C = 12204      XOP =  7919


22 Apr 97    23:37:43    Updated: 04/22/1997          0.382  0.618
Focus Number   File BDWK02        Focus# (High for the swing)  /in 32nds/
Point Number   Support Fib Nodes    Point# (Enter highest reaction low first)
-------------------------------------------------------------------

12204    1
       ===>  11205  *
       ===>  10600  *
9601
-----------------------------------
```

Copyright (c) 1996 CIS, Inc.

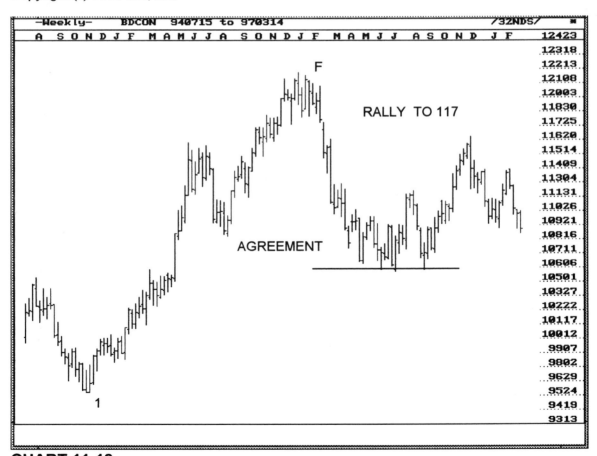

CHART 11-18

HIDDEN D-LEVELS™:

In trading, when everyone knows something, it does no one any good. When *you* know something and everyone else finds out about it later, *is does you a lot of good!* If everyone believes the stock market or soybeans are going up, they are already long and that knowledge is factored into the market. If you know the outcome of a report or otherwise have inside information that is *truly significant*, then you can position yourself ahead of time and benefit from those that follow. Such is the case with hidden D-Levels.

Consider Chart 11-19 below. Based on what you have learned so far, label the Focus, Reaction Numbers, and Fibnodes, before turning the page.

CHART 11-19

Did you include Lineage markings in your labeling? Below is the correct labeling of the chart in question. For clarity, the Fibnodes are not shown.

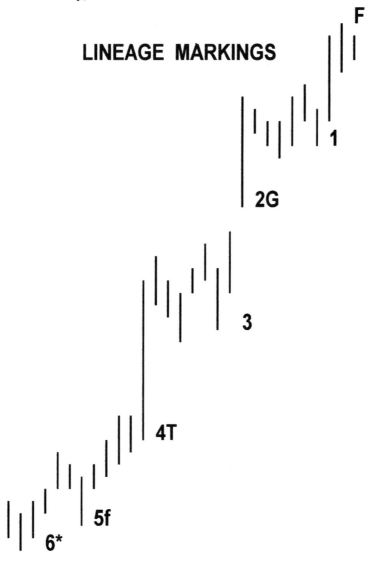

CHART 11-19A

Let's go over this Market Swing one label at a time. The Focus Number is the high of the move. Reaction 1 is the first low preceding the Focus Number. Three bars to the left of Reaction 1 is a slightly lower low, which could have been included in the labeling as 2m (for minor). I have omitted this reaction for clarity, since the Nodes it produces would have almost the same numerical value as those produced by the reaction labeled as Reaction 1.

The second reaction is the top of the gap or the bottom of the bar after the gap. This is a *hidden* Reaction Number and it produces Fibnodes other traders will not be aware of. Fibnode Confluence areas may also be produced which are transparent, even to traders with knowledge of advanced Fibonacci techniques.

Reaction 3 carries no particular Lineage designation. For clarity, I chose it alone, rather than including the Reaction 3 bars to the left of it. Reaction 4 is particularly important since it is off a thrust bar. Like 2G, it produces *hidden* Fibnodes. 4T is more powerful than 2G, since Thrust Reactions are more significant than Gap Reactions. Reaction 5 has the small letter 'f' assigned to it to designate 'first.' This particular reaction is important for its ability to get you in the move There will be more on this in the advanced comments section coming up shortly. Reaction 6 is the Primary Reaction Number designated '*'.

Try labeling Chart 11-20 before turning the page, with your enhanced knowledge of D-Level labeling.

CHART 11-20

This is the proper labeling, considering the hidden Reaction Numbers.

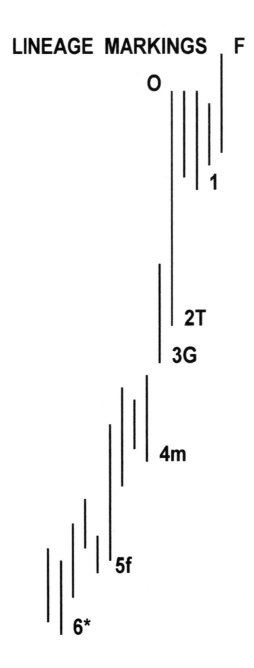

CHART 11-20A

ADVANCED COMMENTS:

If I'm trading an active market like the S&P, I'll usually insert reaction lows, even if they are close to one another, and label the lower one (in an up move) with the letter 'm' to indicate 'minor.' In the course of trading I will enter trades consistent with context of course, on the higher Fibnodes created, thereby insuring a fill. I observe the 'm' Fibnodes to determine the strength of the market, i.e. if I am filled and the 'm' Node is not broken, it indicates more strength than if the 'm' Node is hit or slightly exceeded. Without adequate software, or if you were only using dividers, including these minor Nodes would be inadvisable. The time it takes to calculate them and the clutter this presents, particularly on an intraday basis, would be counterproductive. Lines produced by graphics software splayed across the screen would be equally counterproductive.

Remember earlier when I discussed the difficulty you would face when you bought or sold in the right areas? There is great demand for a fill when you are buying at or near the low point of a dip (DiNapoli, "The X'd Trade".) Using Nodes from a slightly higher Reaction Number, like 1 or 3 on Chart 11-19A can be *invaluable* since you *need to get a fill to make any money!* These Reaction Numbers produce Fibnodes that will enable you to be filled ahead of other traders. Mark the Fibnodes on the chart with your dividers if you don't implicitly understand this. See the more detailed labeling of Chart 11-19 on Chart 11-21.

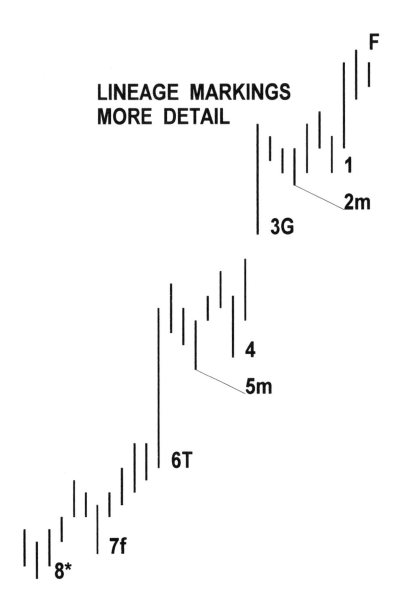

**LINEAGE MARKINGS
MORE DETAIL**

CHART 11-21

This same reasoning is the basis for my use of the 7f Nodes rather than 8* Nodes, upon entering orders. The world may be trying to buy at the .382 of 8*. Who knows about, or considers the .382 of 7f? If you try to be too conservative and buy at the .382 of 8* instead of just above the .382 of 7f, *you're more likely to get filled only when the Node is destined to break!*

Finally, if I can get a Confluence area off of a thrust bar *that is not an identifiable reaction low*, I am an excited trader. I have strong information the world lacks. *I'm looking to make some serious money!*

USING FIBONACCI ANALYSIS TO DEFINE MARKET MOVEMENT:

I've alluded to this technique when I discussed Leading and Lagging Indicators in CHAPTER 2. One of my comments referred to utilizing Leading and Lagging Indicators to their best advantage. Using Fibonacci as an indicator to define Movement might be a bit off the mark. It really depends on your experience level with the market in general and Fibonacci concepts in particular. That being said, the basic idea behind this technique is to look at the extent of the retracements, to determine the anticipated Movement of the Market.

Deep retracements presumably lead to a change in Movement from up to down, for example, while shallow retracements would be consistent with a continuation of existing market Movement.

While I have used and taught this method since the mid 80s, I suggest using it only as a confirming indicator, rather than as a primary Directional Signal or Trend indicator. One has to be *thoroughly schooled* in the application of *higher Time Frame D-Levels* to keep this particular technique accurate.

CHAPTER 12

TYING IT TOGETHER
A BASIC EXAMPLE

Okay, you've studied context, you know whether you want to be long or short. You've studied D-Levels so you know how and where to enter (sort of), and you have a pretty fair idea of how to go about taking an LPO (logical profit objective). Finally, it's time to get a trade on.

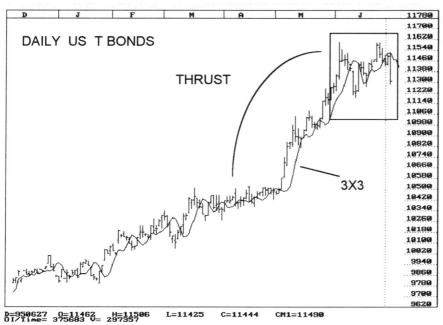

It's June 27th. We have had strong thrust up on the daily bond chart, and a Double RePo has occurred. The two days following the Double RePo show the 3X3 containing the Trend (cupping market action).

CHART 12-1

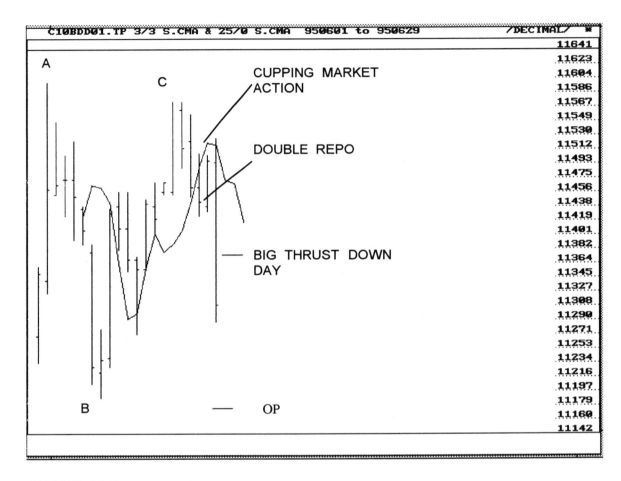

CHART 12-2

Daily-based players could have sold at the market, on close, the day of the Double RePo. An alternative strategy would have been to attempt to sell a retracement back in the direction of the 3X3 the following day. If intraday charting equipment were available, it would be preferable to sell a retracement back on a shorter Time Frame chart. You could, for example, fade a Stochastic buy against a precalculated D-Level on an hourly or a half-hourly chart. A daily-based protective buy stop should be above the '*'.618 Fibnode on close, since that's where the sell signal would fail. To calculate this retracement level, your Focus Number would be on the low of the Double RePo day. Your first (and only) reaction would be the high at Point C. If that type of stop placement is too much for you, a lower stop, behind a daily or hourly D-level would be okay. Just remember *if your stop gets hit intraday and the '*'.618 has not been exceeded on close, then the Double RePo is still in play.* You should resell at the first opportunity.

Once positioned on the short side, you would look for an LPO, consistent with the Time Frame you are trading. Such a calculation on a daily chart is possible using the A, B, C, as shown in Chart 12-2. Note that this (daily) ABC expansion would give OP support just below the low at B. Most traders, unaware of this calculation, would have their sell stops below B, at exactly the wrong place! Also, note that the COP Objective (not shown) has already been fulfilled! It would be prudent to check the Detrended Oscillator to see how oversold we were at that point, before getting too excited about staying with the trade, to the daily based OP.

While the daily is interesting, the shorter term situation is more telling, so that's where we'll focus. Before we get to those details, however, let's first recognize the fear most traders have of losing their position. It's not just the neophytes who think this way. Professionals commonly talk about how hard it is for them to *regain* a position once they have *lost* it, or in our terms, once they have taken a logical profit. What they are really saying is that they don't know how to get into a running market. What I am saying is, even if you lose that "one in 20" or "one in 30" big market move, you're better off taking an LPO, rather than letting the market take you out on a trailing stop. You will see that most of the time, you *can* get back into a running market, or for that matter, initially enter a running market "safely," if you know how and if you do your homework. To illustrate this point, we're going to see how Diligent Dan would have handled a couple of different situations where common knowledge would have reasoned it was *too late*, or the market had gone *too far* to enter. We'll also see how Hyper Hank manages to short circuit his considerable knowledge levels by his lack of psychological control.

SCENARIO 1:

Dan was fat with profits after five weeks of hectic short term trading. Some R&R was in order, so he closed all his positions and took a little trip to Cancun. Upon returning home, the night before the big thrust down day, he found 40 faxes from Hyper Hank describing the Double RePo that had occurred two days previously, and 65 trades he had entered on a one minute chart. Although Hank's faxes accurately indicated that he had 90% winners, he failed to mention that they were for approximately two ticks each. By the time he straightened out that error and paid his broker, Hank had barely broken even.

Diligent Dan seeing the Double RePo and subsequent containment of the Trend by the 3X3, prepared for an entry at the opening the next morning. He attempted to send a fax to Hank three times, but the line was perpetually busy.

CHART 12-3

Above is the first portion of the five minute chart, of the big thrust down day (6/29), shown previously on the daily Chart 12-2.

At the opening, the market gapped down from the night session close. Being unsure of where he'd be filled, Dan waited for a retracement to occur before he entered his orders. We're looking at a down wave so the Focus Number is the low of the move, 11411. The first Reaction Point is the high at 11508. Although we could put a gap 'G' Reaction Number at the high of the day (6/29), let's keep this example simple. We'll consider more advanced comments later. The Fibonacci retracements are shown in the FibNodes™ program printout below.

```
Focus Number   File BDU051        Focus# (Low for the swing)  /in 32nds/0.382 ..618
Point Number   Resistance Fib Nodes   Point# (Enter lowest high first)
-------------------------------------------------------------------------------
11411     1
        ===>  11422
        ===>  11429
11508
------------------------------------
Copyright (c) 1996 CIS, Inc.
```

Dan placed his sell order just below 11422 and his protective buy stop above the .618 Node at 11430. Dan was filled, since there was an almost perfect .382 retracement at 11421. Dan expected a deep move down, since there was a Double RePo in play and the day started with a gap down. Dan therefore quickly calculated the expansions during the consolidation and chose an XOP rather than an OP, or COP to close his trade.

```
7 Aug 96   18:41:45              Updated: 07/24/1996    0.618  1.618
Point Value      Objective Points    File  BDU052      /in 32nds/
--------------------------------------------------------------------
   A = 11508       COP = 11403
   B = 11411        OP = 11324
   C = 11421       XOP = 11306
---------------------------------------
Copyright (c) 1996 CIS, Inc.
```

He placed a closing buy order a few ticks above the XOP level. Dan's protective stop was never approached and the market broke hard. On the drop, the price didn't quite reach the XOP. See Chart 12-4. Whether we assume Dan was filled on his closing buy on *this* drop, or that he had to wait a bit, makes little difference, since the market made the XOP after another throw back rally up to 11326. For the sake of simplicity, let's assume a fill.

Dan called Hank on Hank's private line but couldn't get through. Hank was bitterly arguing with his broker over the fill he'd received on his "at the market" (entry) order. Hank was so worried about missing the big move, he didn't wait for a retracement. The fill was so bad, Hyper Hank was in a rage. He closed "at the market" on the throw back rally where Dan went short. After all, Hank was in a 4 tick loss, and it was so incredibly frustrating! Hank was *FLAT* for the big drop.

What about Conservative Carl?

He saw the Double RePo but wanted to be sure it worked before he entered. When he saw it work by gapping down it scarred him. Carl had purchased some unripend bananas the day before. That was enough risk for him so he sat this one out.

SCENARIO 2:

Both Dan and Hank had a second "safe" opportunity to enter on the short side, after the thrust down move to 11310, when most traders would have considered the market to have "gone too far."

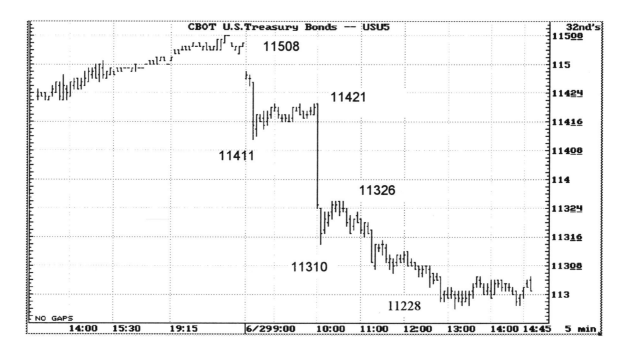

CHART 12-4

Consider the Fibnodes created after the 11310 low. Again, we'll keep this simple by considering only the first reaction back.

```
Focus Number   File BDU053   Focus# (Low for the swing)  /in 32nds/   0.382   0.618
Point Number   Resistance Fib Nodes   Point# (Enter lowest reaction high first)
-----------------------------------------------------------------------------
11310     1
          ===>  11326
          ===>  11405
11421
----------------------------------
Copyright (c) 1996 CIS, Inc
```

If we hit the first Node back, we would be selling at, or below the 11326 level, while hiding our stop behind 11405.

Rather than assume a fill, as we did in the earlier entry, we'll examine a bit of floor mechanics.

Notice the FLAT tops at the 11326, .382 retracement level. This indicates a large overhead sell order or many sell orders.

In this case, an experienced and level-headed trader like Dan, with a resting sell order at 11326, would call the floor and give instructions to "give it a tick." If you can't call the floor you would "cancel-replace" the order a tick or two under 11326 or just "go to the market." In any case, the 11326 area was a nice "safe" place to go short, with your protective buy stop just a few ticks away (over 11405).

ADVANCED COMMENTS:

Let's go over the second entry, Scenario 2, in a more detailed and more advanced level.

The down wave after the initial XOP Profit Objective of 11310 was met, would look something like this on a line chart.

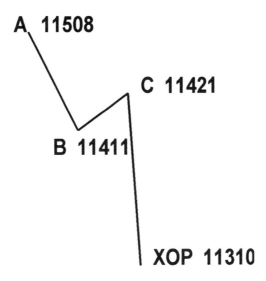

CHART 12-5

The retracement series showing D-Level resistance using the proportional divider, would look like this:

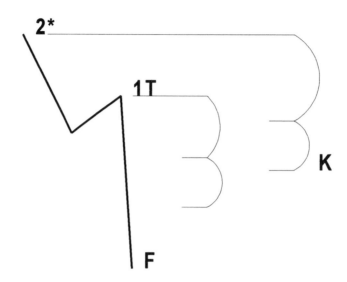

CHART 12-6

The computer printout would display the more complete Fibonacci series and accompanying DiNapoli Levels in a tabular format, that would look like this:

```
Focus Number   File BDU053          Focus# (Low for the swing)  /in 32nds/
Point Number   Resistance Fib Nodes   Point# (Enter lowest reaction high first)
-------------------------------------------------------------------------------
11310    1
        ===>  11326   T
        ===>  11405   T
11421
------------------------------------
11310    2
        ===>  11402   *
        ===>  11416   *
11508
------------------------------------
Copyright (c) 1996 CIS, Inc.
```

The first sell level would be at 11326, and since we are attempting to enter a running market with considerable thrust after a Double RePo, it would be very reasonable to get aggressive and hit (sell) the first DiNapoli Level calculated. As mentioned earlier, if a sell were attempted at 11326, given the flat tops, it would be prudent to call the floor and tell them to "give it a tick." In any case, selling a tick or two before the Node is reasonable, otherwise you are apt to be filled, *only when you are wrong, i.e. when the trade goes against you!*

Sometimes it takes more than good analysis to be a winner and in this case a knowledge of market mechanics is essential. What about selling at 11326 MIT? "Market If Touched" (MIT) orders are typically not accepted at the Board of Trade where bonds are traded. You can always make other arrangements if you trade enough, but if you're a "two lotter" without access, and you don't give it a tick or two, you may not be able to gain access to the floor in time to change your order.

A good stop for this trade could be behind the 'K' Confluence area, i.e. 11402-11405. This Confluence area is made up of a 'T' (Thrust) .618 Reaction and a '*' (Primary) Reaction .382 Node. It doesn't get any better than that. As strong as this area of Confluence is, however, I would place my *initial* stop behind the major (11416) '*' .618 instead. Here's why. If the market pushed through 'K' briefly (a few seconds), a stop placed at 'K' would be hit, but a stop placed behind the '*' wouldn't be. A brief push through Confluence *does not* mean the area is broken, so you would want to stay with the trade.

If, however, the market stayed above 'K,' I would suspect my positioning in the trade was wrong, and take the first opportunity to get out on a pull back toward the Confluence area. Then I'd cancel the initial stop.

After that, I'd take a fresh look at the trade by examining the expansions up (seeing if they were fulfilled) and the current Trends on the intraday charts to see where I was, i.e. whether or not the context for the trade supported further action.

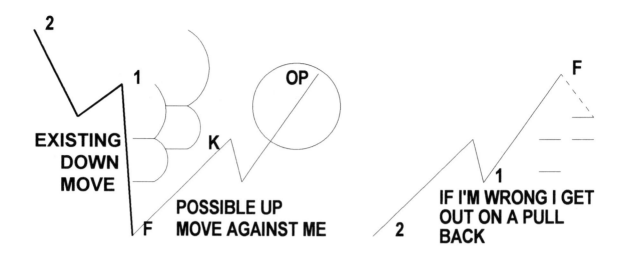

CHART 12-7

Now let's consider in more detail the possibility of the market moving against me as shown in the left portion of Chart 12-7. We have the market breaking back down after being held at Confluence, then supporting on a Fibnode, and making a move up against me to an OP. In this case, I have a couple of options. The re-test of the Confluence area may result in a double top and a subsequent break to new lows. However, it would be prudent to lower my buy stop from its initial position, to the high from 'F' to 'K.' Stops now exist above this most recent high. Therefore, there is a higher probability that the Confluence area will be penetrated on a *second* push up, than when it was initially approached. If I were trading size or if I were in a thin market and I were *very concerned with fills* resulting from stop orders, I could wait for a retracement of the full up move and get out on an "Or Better" (OB) order against a support Fibnode. This possibility is pictured on the right side of Chart 12-7.

If this level of detail in the "Advanced Comments" section has you reaching for the Advil®, just ignore it. If you can't figure out why anyone would bother with just a few ticks, the answer is threefold. First, a few ticks on a 100 lot *is significant* in the bonds. Second, these advanced comments can be applied to any Time Frame, not just to this example, so they are worth considering as an intellectual exercise. Third, if you trade size, you're often in a much better situation closing your position with an OB (or better), rather than stop orders.

NOW LET'S GET BACK TO REALITY:

We can reasonably assume Dan achieved a fill, selling at 11325, when he told the floor to "give it a tick." We can see why he took the COP out on all or a part of the position he sold, when we consider the comments below. .Here's what the expansion series looked like.

```
24 Jul 96      12:22:12           Updated: 07/24/1996   0.618  1.1618
Point Value     Objective Points     File   BDU054       /in 32nds/
-------------------------------------------------------------------
   A =  11421      COP =  11231
   B =  11310       OP =  11215
   C =  11326      XOP =  11120
-----------------------------------------------
 Copyright (c) 1996 CIS, Inc.
```

The reason I say "all or part," is that on the one hand, it's a Double RePo and is likely to go lower over time. How soon, may depend upon how oversold the market is at the COP level. As I suggested at the beginning of the chapter, look at the detrend to determine this, or use the Oscillator Predictor™ to determine it ahead of time. On the other hand, I always like to book logical profits, and it's been a big day. Also, if we are strictly intraday players, it's getting late, and we would probably be expecting a lot to be looking for an OP *before the close.* As it turned out, the COP expansion 11231 was met and exceeded slightly, so we would have had no problem achieving a closing fill.

I've contrasted the actions of a level-headed trader and an emotional trader to show you how personality traits can impact performance. It's obvious that there are alternative trading strategies, outlined earlier in the book, that meet the challenges illustrated in this example. The implementation of these alternative strategies depend upon the individual trader's experience, access, Time Frame, and personal objectives.

FREQUENTLY ASKED QUESTIONS:

Did Hyper ever get a decent trade off?

No, Hank never got his fill, selling at 11326. He made arrangements to change brokers and spent the rest of the day at the local bar drowning his frustrations and talking to anyone who would listen. He never had a chance. He was exhausted before the day began and impotent when he most needed focus. Now he'll be out of the market for weeks, while he waits for his new broker to mail him the correct forms and for his funds to be transferred. Chances are he'll need a new bite plate before he gets over this one.

Joe, you say you can reenter a running market, but we both know you'll never get a fill on a big drop, so how do you enter in that case?

The problem is one of semantics. You enter a running market against a D-Level retracement with an appropriate stop. If it's a disaster situation, however, like October '87 in the S&P, you don't play[1]. Risk tolerance is good for trading. Foolhardiness is suicidal. When I see a market in panic, I am not interested and I look for a way out as an upcoming S&P example will illustrate.

I understand the example in its entirety but, according to the criteria for a Double RePo in Chapter 6, it seems the length of the first and second penetration are a little too wide. Is this a look-alike or a Double RePo?

Strictly speaking, it's a look-alike. Given the extent of the preceding thrust however, I have no problem with treating this look-alike as the real thing.

---

[1] Reference the in home TRADING COURSE (Andre the Giant and Chuck Norris anecdote).

CHAPTER **13**

FIBONACCI TACTICS

ᴏᴏ

GENERAL DISCUSSION:

Various Fibonacci-related trading strategies have been discussed throughout this book. Some have catchy names designed to help you remember their character, some don't. What qualifies as a trading tactic and what might be considered a trading hint is a debate I won't get into. Regardless of what these strategies are called, I believe it's particularly helpful for you to have a chapter devoted specifically to this topic, and to have certain of these approaches defined, apart from market examples. Those of you who have studied my earlier material may recognize that fewer tactics have been described in this chapter than you have seen before. The reason is simple. Although *all* of the tactics described earlier are effective and continue to work, I believe that not all of them deserve equal consideration.

Now, before we address the specific techniques described, let's be sure we understand the problems they are designed to solve.

Consider the following Fibonacci support series depicted in the FibNodes™ program printout.

```
Focus Number   File SMNDA4        Focus# (High for the swing)
Point Number   Support Fib Nodes  Point# (Enter highest reaction low first)
-------------------------------------------------------------------------
17760    1
      ===>  17416  M
      ===>  17204  M  ───────┐
16860                        │
----------------------------------      K
17760    2                   │
      ===>  17214  f  ────────┘
      ===>  16876  f
16330
----------------------------------
17760    3
      ===>  17080  *
      ===>  16660  *
15980
----------------------------------
Copyright (c) 1996 CIS, Inc.
```

This series clearly shows a Confluence area between 17214 and 17204. We will assume some appropriate context that indicates a choice of that Confluence area as our long entry point. We'll choose 17220 as our specific entry price and call that point 'X.' The assumed context could be something previously discussed in this book or something you have chosen as an appropriate indicator. For example, maybe you want to look at put-call ratios, Bollinger Band extremes (on the pullback), a respected recommendation, a Commitment of Traders report, or whatever. A deeper D-Level, let's say just below the 17080 level, will be the protective stop placement area 'Z.'

Problem: How can we best enter this market at 'X,' where we *assume* support will manifest?

Answer: Different strokes for different folks, hence there is more than one Fibonacci entry tactic.

NOTE:

Be sure you understand how I have framed the forthcoming discussion, before you go on. The next series of important examples all hinge on the preceding paragraph.

BONSAI: AN ENTRY AND STOP PLACEMENT TECHNIQUE

Although I rarely use this particular tactic, I include it for those of you who are psychologically suited to it. This would include traders who are willing to accept a lower percentage of winning trades than would be possible using other techniques described in this book. A number of former floor traders I have trained like this technique. Hyper Hank would likely be a Bonsai enthusiast. Those who utilize Bonsai believe, perhaps correctly, that their ultimate level of profitability will be enhanced by this entry tactic, since they typically trade more often and keep the level of each loss suffered very low. Here's how this very simple strategy works.

You have, as indicated above, a pre-calculated D-Level entry at point 'X.' Using Bonsai, you have a preset money or point stop at 'Y,' *irrespective of additional DiNapoli Levels.*

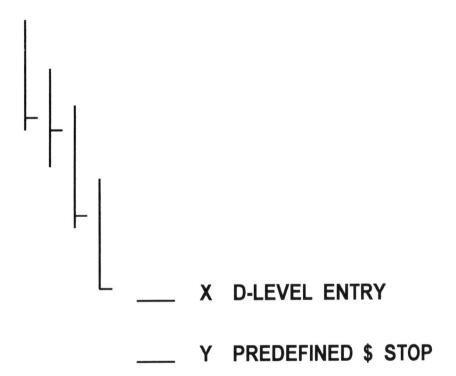

X   D-LEVEL ENTRY

Y   PREDEFINED $ STOP

CHART 13-1

You enter both orders at the same time and hope the stop isn't hit. If it is, and the price immediately gets back above 'X,' you then reenter the trade at the market and again place your stop at 'Y.'

If the original stop at 'Y' is hit and the price remains below 'X,' then see if the context you choose to support the trade is still valid. If it is, you select a deeper DiNapoli Level upon which to enter the trade and place another money stop below it.

Bonsai players have 'Y' defined differently for each different Time Frame they may choose to play. The amount is usually determined by individual experience in a given market. Between 55 to 85 points is very typical for the five minute S&P, while three to five thirty-seconds (3-5/32nds) is common for the five minute US Bond market. The advantage of using Bonsai is its simplicity. Its ease of use frees the mind from more complex stop loss exit strategies and therefore may preserve a more relaxed attitude in the trader. He is therefore free to properly pursue the next trade in this, or other markets.

There are many disadvantages. Bonsai players typically disregard volatility from one day to the next in their selection of pre-set money (or point) stops. If they don't have excellent brokerage services, the frequency of trading can be costly in both slippage and transaction fees. They also must have quick floor access to reenter orders if they are trading short term. Although the exit criteria is simple, trades require constant observation, since contingency order criteria left with a typical broker would likely be too complicated and subjective for the broker to be held responsible for the result.

BUSHES: AN ENTRY AND STOP PLACEMENT TECHNIQUE

Bushes is the technique typically used in the examples in this book, with the exception of those described in "Advanced Comments." .You enter your buy above one D-Level and hide your stop behind another. It gained its name from a market professional who attended a private seminar some years back. He crouched behind a large plant I had in my office area. Then, as he shot at another attendee with his finger, he indicated that his favored method to capture profit was to hide behind the bushes, jump up, take a shot, and duck back down behind the bushes. The entry and stop strategies described in CHAPTER 11 (AN IDEALIZED TRADING EXAMPLE) are clear examples of Bushes. Standing up and taking the shot is akin to your entry before one D-Level, while crouching down behind the bush is akin to hiding your stop behind another D-Level. The differing trade entry strategies described in CHAPTER 11 are simply variations in where to take the shot and which bush to hide behind.

The obvious significant difference between Bonsai and Bushes is the lack of "cover" for the stop. The subtle, significant difference requires a knowledge of floor mechanics. Suffice it to say, if your stop is just behind a support D-Level, you have a good broker, and you're hit, slippage should be mitigated. If you have no support (more likely with Bonsai) even with a good broker, who knows where your stop will be filled. Your chance of having a better fill is always higher with Bushes than with Bonsai because of the support manifested by the D-Level itself, even if it doesn't hold.

                    X   D-LEVEL ENTRY

                    Z   D-LEVEL STOP

CHART 13-2

MINESWEEPER A: AN ENTRY AND STOP PLACEMENT TECHNIQUE

Using this technique, we're exercising more caution than with either Bonsai or Bushes.
Here's how it works. Let's say we want to enter the market at D- Level 'X.'

**D-LEVEL
ENTRY X**

F

1

CHART 13-3

While we expect support at 'X,' we wait, first for support to manifest at Point 1, then for
a move up to 'F.' Point 1 and Point F (our Focus Number) are selected by market forces.
We do not attempt to precalculate them. After those occurrences, we calculate the Fib
levels of that up move. On a line chart, it would look like Chart 13-4.

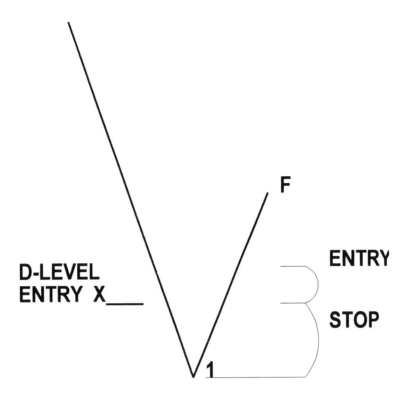

CHART 13-4

Our actual entry would be above the .382 Node. Our stop would go under the .618 Node or against the old low at 1. We've bought insurance which may or may not be costly depending on what market action actually develops around the selected D-Level entry 'X.' With the benefit of hindsight, we can explore a variety of possibilities.

POSSIBILITY 1:

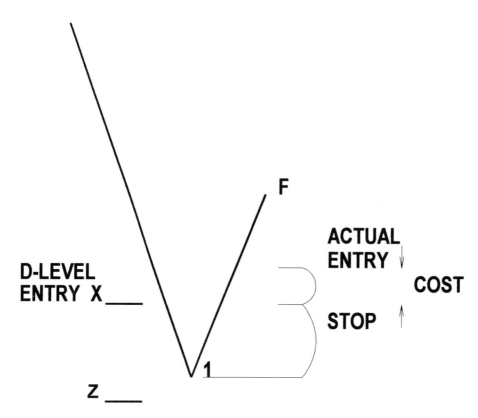

CHART 13-5

If we had support very near 'X,' our stop at 'Z' wouldn't have been hit and our actual entry would have been higher than an entry at 'X.' Our insurance policy in this case would be *costly* as represented by the distance between the arrows on Chart 13-5. Don't misinterpret this. The cost is between our original D-Level choice 'X' and our actual entry, not between our actual entry and our stop. Of course, we might not get the fill in the form of an expected retracement back to our actual entry, but it's likely we will.

POSSIBILITY 2:

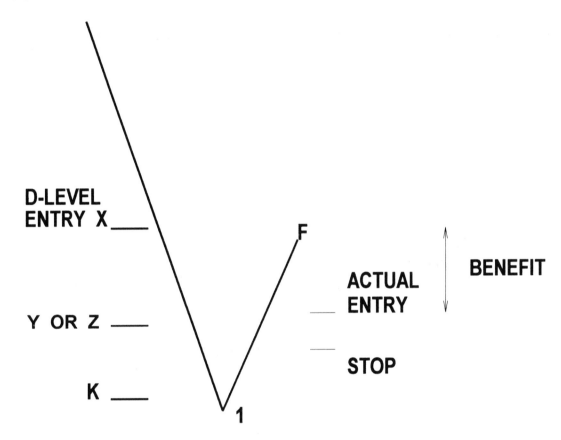

SOME BENEFITS

CHART 13-6

Let's assume the expected support area at 'X' was deeply penetrated, all the way through 'Z,' to a deeper Node or Confluence area 'K.' 'Y' in Chart 13-6 represents a Bonsai money stop. It's included here, as well as the Bushes stop, so the Bonsai players among you can see the effects of possibility 2. In this scenario, we would have been stopped out if utilizing the Bonsai or Bushes entry tactic. Waiting for the Minesweeper entry prevented getting stopped out. Additional advantage was achieved since we entered at a lower price than the original D-Level location at 'X.' So we achieved some significant benefits utilizing this tactic under these conditions.

POSSIBILITY 3:

If we essentially had no support, as in an illiquid market, then we avoid a big ouch by utilizing the Minesweeper entry!

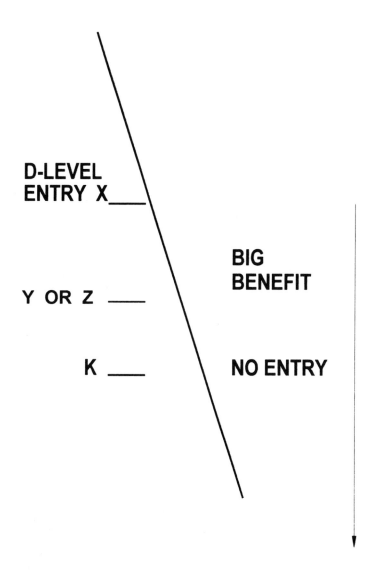

SIGNIFICANT BENEFIT

CHART 13-7

MINESWEEPER B: AN ENTRY AND STOP PLACEMENT TECHNIQUE

This technique attempts to buy *more insurance* by placing the actual entry *above a Confluence area*, rather than simply above a Node. The stop would have a greater degree of protection since it is located below a Confluence area. You might say the bush is larger and thicker. This is what a line chart of Minesweeper B would look like. You have several choices of entry, above the first .382 Node, above Confluence, or even above the Primary .618 Node (not shown.) You have choices for your stop as well. Below Confluence, below the .618 Node, or against the low at Point 1.

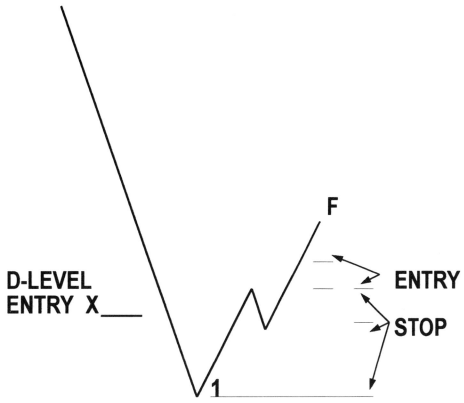

CHART 13-8

Both Minesweeper A & B are typically employed by dropping the Time Frame to achieve more reaction lows from which to calculate the DiNapoli levels, *after support is manifest.* This is a good example of why it is so helpful for daily-based players to have access to hourly charts, even if they are developed from a delayed data feed. See CHAPTER 1.

ENTRY TACTICS USED ON AN HOURLY S&P:

The following example shows how some of the tactics look when applied to an hourly S&P.

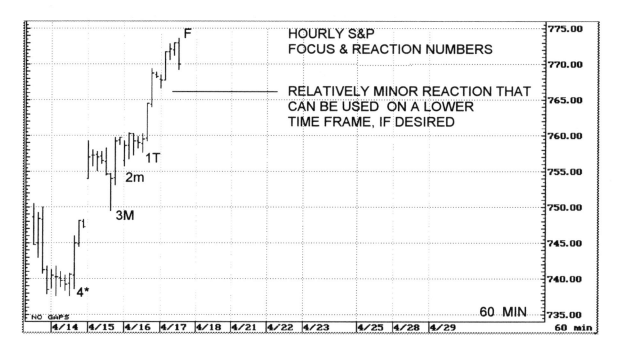

CHART 13-9

In Chart 13-9, we have a strong up move in the hourly S&P. We'll assume the daily and hourly Trend is up and that we'd like to get long this market. We'll also assume a highly overbought level at 70%, so we don't want to get in too quickly and once we're in, we don't want to get "married" to the trade. Our final assumption is that the higher Time Frame Fibnode resistance locations will not affect our trade, i.e. they are *not in play* within and somewhat beyond the price range shown on this chart.

The Fib series printout below reflects the Focus and Reaction Numbers shown on Chart 13-9. I have not shown the Confluence area, since by now you should be able to locate it on your own by comparing top and bottom Nodes.

```
Focus Number   File SPM602        Focus# (High for the swing)
Point Number   Support Fib Nodes     Point# (Enter highest reaction low first)
------------------------------------------------------------------------
77360     1
     ===>  76812   T
     ===>  76473   T
75925
---------------------------------
77360     2
     ===>  76678   m
     ===>  76257   m
75575
---------------------------------
77360     3
     ===>  76439   M
     ===>  75871   M
74950
---------------------------------
77360     4
     ===>  75985   *
     ===>  75135   *
73760
---------------------------------
```
Copyright (c) 1996 CIS, Inc.

Okay Hank, where do you get long?

Hyper Hank: *Before I answer that, I'd like to know why we don't have the first reaction about four bars to the left of the Focus Number included in here. It's valid, isn't it?*

Yes, it's a valid Reaction Number, but the trade criteria was that we were overbought. We are *not* looking for the highest possible entry level (See CHAPTER 7). If we were looking for a higher entry level, and we had the capacity, we'd drop our Time Frame to a 30 minute or a five minute chart. This reaction and others that we can't see on the hourly, would be included in a shorter Time Frame series.

Hyper Hank: *Okay, Okay, I've got it. Well, I'd enter above 76812 and put my stop below Confluence at 76473. But I don't like that as a stop.*

Why not?

Hyper Hank: *It's over a 300 point stop; I can only trade one contract!*

What would you really like to do?

Hyper Hank: *I'd like to find out where the Fibnodes are to that first reaction and put my stop under one of them.*

What do you say Dan?

Diligent Dan: *I'll enter above Confluence at 76480. Then I'll use a Bushes stop below the lower end of Confluence at 76425.*

Why?

Diligent Dan: *If the trade is good, Confluence will probably hold and the next Bushes stop is under 76257. That's a minor Fibnode, so I'm not excited about using it.*

What's Carl going to do?

Conservative Carl: *Since we are so overbought, I'll wait to enter the lower Confluence area between 75871 and 75985. I'll buy it at 75875 and use a Bushes stop below the 75135 Primary Node. I know it's a big stop, but I've made an arrangement with my friend who's afraid to trade the S&P alone, to take half of my S&P trades. With a half of a 1 lot, the stop isn't too bad.*

Have you had him deposit money into your account?

Conservative Carl: *No, but I'm sure he'll pay me if something goes wrong.*

HMMM...

Let's see how the trade worked out on Chart 13-10.

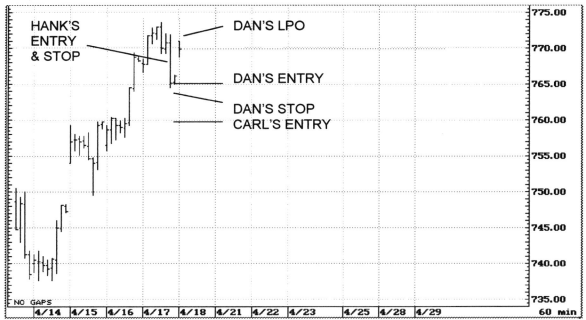

CHART 13-10

Hank saw the market just miss his entry so without telling anyone, later that day he canceled his entry and went in at the market on a 10 lot (not shown because he didn't tell us.) He put a Bonsai 55 point stop under his entry level. He got both fills back at the same time. The floor loves Hank. So does his dentist, he went through another bite plate.

Dan got his fill and his stop was never hit. He opted for the .618 Primary resistance Node as his LPO, since the market was overbought. Because his closing order was in the market *prior* to the next day's open, he did even a little better than he expected.

Carl was too nervous to watch the market after the big thrust down bar that filled Dan's order. He was very relieved when he came back from tending the garden, to see that he was never filled.

THE TRADE CONTINUES (Chart 13-11):

On the second dip down, Dan waited for a Minesweeper entry after the low was formed at 758.30. He did this for two reasons. Dan saw the thrust down on the hourly chart and it made him nervous, since it was the *second* time down after up thrust. Although he didn't use the 3X3 on an hourly chart, this formation was looking a bit like a Double RePo sell. He also could not discern a reasonable stop level.

Carl left his orders as they were, reasoning that nothing significant had changed to impact the original trade criteria. The daily Trends were still in his favor!

Hank was not playing this one. While he watched the screen, he was once again busily calling relatives and friends trying to raise capital.

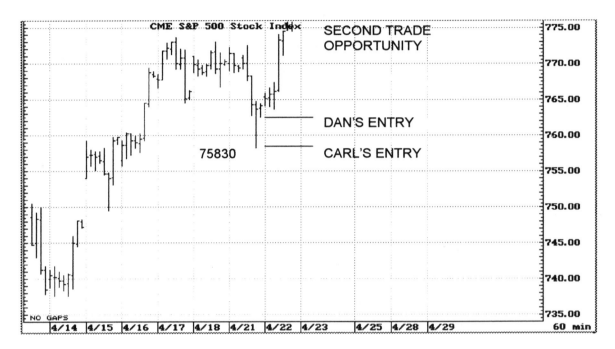

CHART 13-11

Let's see how this trade worked out:

Carl was filled and his stop was never hit. He and his partner took an almost perfect OP out on the trade (See Chart 13-13).

Dan's Minesweeper entry was filled and his stop against the preceding low was never hit. He noticed the RR track about an hour after his initial entry and doubled up his position by recalculating the Fibnodes and entering more long positions at the .382 Node. He blew out all his positions on the same OP in Chart 13-13 that Carl got out on. Chart 13-12 shows his entry and stop points. Upon reexamination of his trade criteria, Dan realized he had missed the Agreement between the Confluence area Carl and he had gone long against and the XOP of the down move, as shown on Chart 13-12. He wasn't concerned about it, however, since he had just booked $12,000 per contract in two days. He needed some R&R and was off to another vacation, this time to Bangkok. After making some notes about the trade, which he always did, he was off to catch a plane.

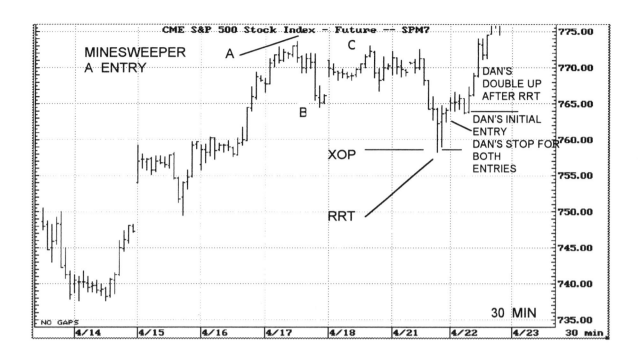

CHART 13-12

CHART 13-13

Objective points for Chart 13-13:

```
30 Apr 97  13:44:35           Updated: 00/00/0000    0.618  1.618
Point Value      Objective Points    File  SPM301
--------------------------------------------------------------------------

    A = 74950        COP = 77319
    B = 77360         OP = 78240
    C = 75830        XOP = 79729
```

Copyright (c) 1996 CIS, Inc.

WASH AND RINSE: A CONFIDENCE BUILDER

This pattern is more in the category of a hint or a clue than a tactic. Let's say you have consolidation above a DiNapoli Level but the market has refused to get in gear and move to the upside. Then, suddenly, all stops have been cleaned up (taken out) and on a lower Time Frame chart, you begin to get some thrust. The fact that the D-Level has been briefly penetrated is not of concern.

There are a variety of examples showing the Wash and Rinse phenomenon in this book. Some are shown in Chapter 15. The Wash and Rinse does not have to occur at a D-Level, it's just easier to play if it does.

Why does this work? The idea here is that once the market has been cleaned up, the floor has no incentive to go lower. Another possibility is that a search of lower levels by market participation has found substantial buying. A third more cynical interpretation would indicate that whatever entity, or trader(s) that wished to accumulate a significant position is now satisfied (by buying the consolidation as well as the sell stops). This entity or individual, now satisfied, has no incentive to hold the market back by acting to force a consolidation area to be formed. In any case, all those who were stopped out, who wish to participate, must reenter at higher levels. All intraday Trend indicators start to point up and it's time to get involved. If you exited[1] or were stopped out, it's time to reenter, by using one of the entry techniques described above.

---

[1] DiNapoli, "Three Period Rule," *FIBONACCI MONEY MANAGEMENT AND TREND ANALYSIS in home trading course.*

FREQUENTLY ASKED QUESTIONS:

Which technique is best? What should I use?

That's up to the psychological make up of each trader. There is no single answer. When you become thoroughly familiar with each tactic by observing market action and the consequences that each action would have created, you will inherently choose one or the other. Experience in a given market situation will also tend to dictate the appropriate strategy.

Most of my trades are variations on Bushes. I use the other strategies when they seem more appropriate.

Minesweeper B seems complicated. You give up a lot before you get in, so why even bother with it?

It is more complicated but still relatively easy to employ, particularly if the context, or set up, is on longer Time Frame charts, say hourly and above. Whether or not it is costly depends upon how much support develops on, or around the selected entry Node. Refer to the Minesweeper section earlier in this chapter if you don't understand this point.

Isn't the Minesweeper technique just a variation of Bushes, after you see support where you initially wanted to enter?

Yes! I've attached a different name to this strategy and treat it differently to illustrate its applicability.

While all the examples above are shown for long entries, they work equally well on the short side.

CHAPTER 14

AVOIDING A TYPICAL MISTAKE

ᵈᵈ

GENERAL DISCUSSION:

Trading requires constant and focused vigilance. The professional distinguishes himself by making Mistakes less often than the neophyte. He cannot hope to avoid them altogether. By studying this next example, see if you can avoid misinterpreting a situation similar to the one depicted and in so doing avoid some unnecessary losses.

This example is based on monthly data, however the same thought process it imposes can be applied to an hourly or a five minute chart.

YEARLY BOND EXAMPLE:

Consider the labeling of the following US bond monthly continuation chart and the accompanying FibNodes™ printout, which shows three reaction lows and a Focus Number of 12210.

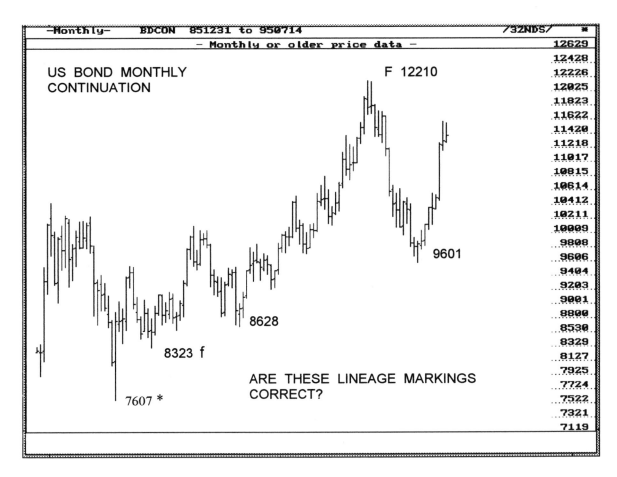

CHART 14-1

28 Apr 97   11:34:11   Updated: 04/28/1997   0.382 0.618
Focus Number   File BDMOPA2   Focus# (High for the swing) /in 32nds/
Point Number   Support Fibnodes   Point# (Enter highest reaction low first)

```
--------------------------------------------------------------------------
12210    1
     ===> 10825
     ===> 10013
8628
----------------------------------
12210    2
     ===> 10718  f
     ===> 9815   f
8323
----------------------------------
12210    3
     ===> 10423  *
     ===> 9326   *
7607
----------------------------------
```
Copyright (c) 1996 CIS, Inc.

We will be focusing on price action in the vicinity of the .382 Primary Node to see what we might learn about future price Movement. However, if you had been trading bonds from higher levels and from a shorter Time Frame chart, it would have been reasonable to assume significant D-Level support at *all* of the support Nodes shown. This is because all of these Nodes were created from important *monthly* reactions, not from a half hour chart. If you look at the daily and weekly charts, you'll find at a minimum, respectable rallies in the vicinity of *all* of these price levels. Consider how this would have helped you, even on a five minute chart!

When we're analyzing whether or not a long term bull market is still intact, we look to our trend indicators on the weekly and monthly charts. We can also observe whether or not the Primary '*'.382 Node has been broken or if a major Confluence area has been penetrated. If major support has been broken, it obviously signals more weakness than if it holds. Such a break could indicate a major change in the Movement of the market. By looking at the monthly D-Level series shown in File BDMOPA2, 10423 *appears* to be the Primary .382 Node. In fact, it is a Fibnode generated from a Major Reaction, not the Primary Reaction low, as we will see. While the penetration of 10423 correctly forecasted lower prices, it is not nearly as serious as a break of the Node created from the Primary Reaction low.

When 10423 was penetrated, I expected a move down to the vicinity of 10013, since that was the next Fibnode support. (See the .618 retracement in Box 1). A reasonable long entry (if supported by context and an appropriate entry tactic) would be in the vicinity of 10013. A reasonable stop placement area *appears* to be just under the .618 Fibnode at 9815, in Box 2. It would also seem reasonable to expect that closes below 9815 for a significant period of time, would indicate much lower prices, perhaps a move down to the last Fibnode support shown in Box 3, all the way down to 9326!

While all this *seems* to be good analysis, an error is being made. What's on the screen, i.e. what we see on Chart 14-1, is *not* representative of what we need for adequate analysis.

Consider Chart 14-2 of the *same* US bond continuous contract. In this depiction, reaction lows back to 1981 are included.

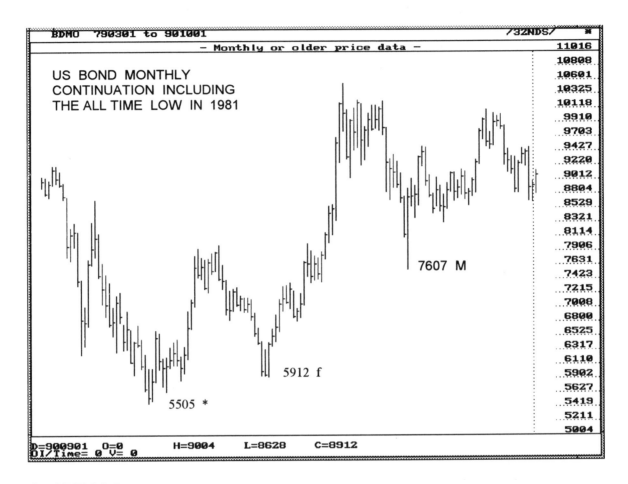

CHART 14-2

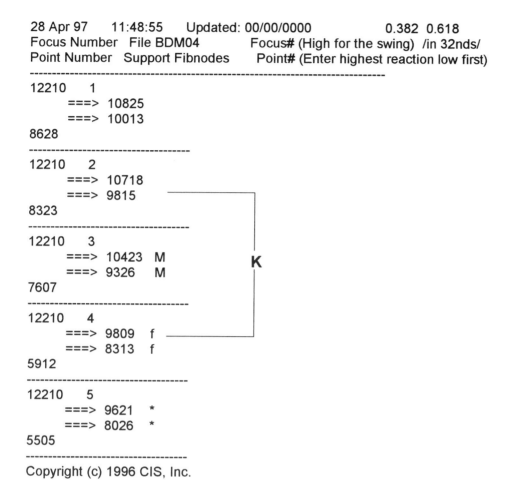

When we look at the D-Level series printout that includes these reaction lows, we get a substantially different picture. Now we see that 9621 is the '*'.382 Node, not 9815! We can also see monthly Confluence between 9809 and 9815. This more complete picture and the accompanying Fibnode series has changed the analysis significantly. For one thing, the last place we would want our stop would be at, or above, the Confluence range of 9809 to 9815. We would certainly want to get below that and we're not looking for 9326 with the Primary Node of 9621 coming in substantially before that!

Now, on Chart 14-1 notice the market action around both the Confluence area, and the .382 Primary Node at 9621. Price came down to the vicinity of the 9815 level, supported, then broke to 9601 briefly, thus exceeding both the Confluence area and the '*'.382 Node. Since this is a monthly chart, "briefly" differs from what "briefly" would mean on a five minute chart. The Confluence area was only *temporarily* broken and the monthly continuation *never closed* below the 9815 Fibnode! The bond market merely reacted back to the '*'.382 Node, i.e. the Fibnode created from the lowest low recorded in the history of the US bonds futures. Our conclusion, based on the proper analysis, can confidently be that the bull market remains intact and as we can see, prices returned to the old high at 122 in just a few months.

Admittedly, it's easy to *say* "merely reacted" back to the '*' .382 Node. If you were holding 50 contracts of bonds on the long side, it wouldn't have seemed all that "mere." What you need to understand, however, is that all this analysis *must be relative to the Time Frame you are analyzing*. Observing this type of action in the monthly continuation chart can produce an excellent context for your shorter Time Frame trades!

On a five minute chart, I *like* to see a major retracement back to the Primary Node. It cleans up the market and allows for an orderly move back up, perhaps to higher prices. It's no different in the higher Time Frames. I have to be sure, however, that what I think is the Primary Node is in fact the Primary Node for the Market Swing I am analyzing.

THE BROADER PICTURE:

This type of analysis is not confined to the typical financial markets we trade daily. When Orange County, California, real estate prices crashed from their highs around 1990, everyone was ready to throw in the towel. Orange County itself went into bankruptcy a few years later. Chicken Little was screaming!

Let's look at an admittedly unscientific but telling D-Level analysis. For ground rules we'll keep certain things constant, like the same type and class of home and the same general neighborhood. We'll also consider a class of home that has always had reasonably good liquidity characteristics. We're not talking lumber contracts here. We're talking the real estate equivalent of T-bonds, your 1200-1500 square foot, two or three bedroom, middle-class, love cottage.

In the early 70s, home values approximated the cost of building which was fairly constant. Inflation was reasonable. Things really started to move with values traveling almost vertically and uninterrupted into the Carter "tight money days" of the early 80s. Our generic love cottage preceding the 20% prime interest rate and 55 on the US 30 year bond futures, was about $115,000. From that high, our cottage reacted back to about $90,000. When interest rates came back within reason, the move up resumed reaching a high of about $215,000, a bit above an OP extension.

Chart 14-3 shows the ultra-simple D-Level yearly series on this slice of the southern California real estate market. With more data, we could include more reaction lows. But, this is a yearly chart and we're only looking for the *major* Fibnodes.

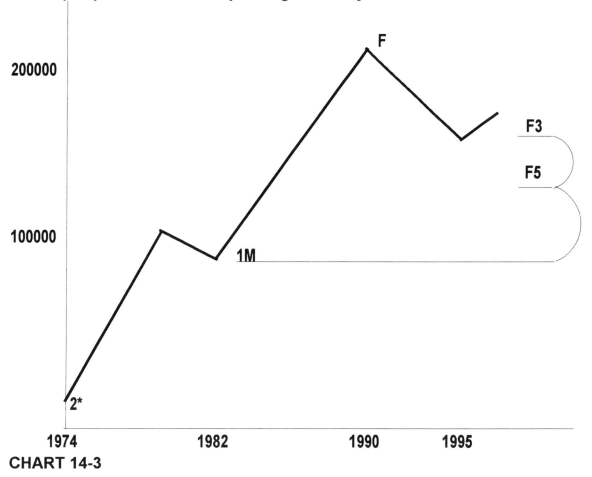

CHART 14-3

28 Apr 97   14:16:59   Updated: 04/28/1997   0.382 0.618
Focus Number   File CAREYR02   Focus# (High for the swing)
Point Number   Support Fibnodes   Point# (Enter highest reaction low first) [omit edit again, sorry]

215000   1
         ===> 167250 M
         ===> 137750 M
90000

215000   2
         ===> 142420 *
         ===>   97580 *
25000

The reaction of price back to $170,000 in the early to mid 90s, *only met the major .382 Node*. Values stayed considerably *above* the .382 primary Node at $142,420. Where's the problem? Throughout the early '90s, I encouraged friends, relations, and business associates to go long Orange County real estate. Very few did - they could not see the support so apparent to someone schooled in Fibonacci Retracement analysis. This down move was just a *normal* correction! Those who took the speculation have some solid equity now!

It's interesting to note that the low was supported not only by the Node off the Major Reaction, but also by rental income - in other words, what the house could earn to pay a typical mortgage. When farm land crashed from inflation's highs, it also returned to an economically viable level, i.e. what the land could reasonably expect to produce. I didn't do the analysis, but I would expect the price decreased to the vicinity of the Primary Node.

Where do we go from here? The reaction down to $170,000 has given us a new place from which to calculate an expansion. I'll leave it to you to calculate the OP, but that's where I think we're going, *if* we can overcome the resistance Nodes (not shown), formed from the "recent" down leg.

While we're on the subject of long term charts and Major Reactions within the context of a continuing up trend, let's see the example of the Dow Jones Industrials experiencing a 500 point, one day drop and a multiple week drop of over 1,000 points. Consider the following D-Level series shown in the FibNodes™ printout on the next page and first depicted graphically in CHAPTER 11.

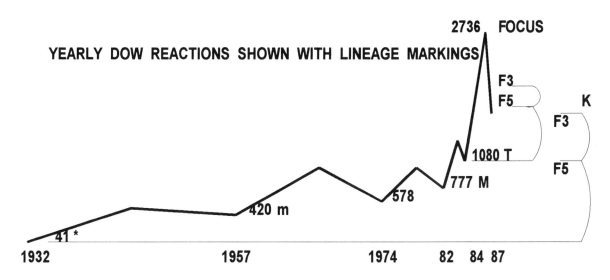

YEARLY DOW REACTIONS SHOWN WITH LINEAGE MARKINGS

CHART 14-4

22 Apr 97   15:27:24   Updated: 04/22/1997   0.382   0.618
Focus Number   File DJYR02   Focus# (High for the swing)
Point Number   Support Fibnodes   Point# (Enter highest reaction low first)forget the edit]

```
2736    1
      ===> 2103   T
      ===> 1713   T
1080

------------------------------------
2736    2
      ===> 1988   M
      ===> 1525   M
777

------------------------------------
2736    3
      ===> 1912
      ===> 1402
578

------------------------------------
2736    4
      ===> 1851   m
      ===> 1305   m
420

------------------------------------
2736    5
      ===> 1707   *
      ===> 1070   *
41

------------------------------------
```

Copyright (c) 1996 CIS, Inc.

We just reacted back to the '*' .382 Node, which was also a Confluence area! *Price stayed within the context of a normal correction.*

A QUESTION

Joe, you can always go back in time and get lower reaction lows! So what's a Primary Reaction and what isn't?

Part of the answer to your question entails bookkeeping or organization and part of it is understanding how Time Frame(s) impact your trading decisions.

While it is *not true* that you can *always* find lower reaction lows by going back further in time (the bonds saw their lows in 1981), it *is* true that by looking at larger Time Frames or compressing your data, it may be possible to locate deeper retracements than shown on your (default) screen. Here's how to handle this situation. Pre-calculate a Fibnode series for all Time Frames above those you are dealing with, as well as the one you are dealing with. If you are an hourly player for example, have the daily, weekly, and monthly series organized. That means have a printed copy attached to a clipboard, labeled for that instrument, and readily available. The charts should be appropriately marked by use of the divider. You should also have these series as computer files in the software you're using. (This software should have a "paging" feature to keep your files properly organized). *Each file, as well as each printed chart, can have its own Lineage Markings independent of the higher Time Frame files.* In our example of the bonds, I would have left the original FibNodes™ printout BDMOPA2 as it was and I would have had a quarterly or yearly file, with its *own* Primary and First Reaction Lineage Markings. The reason that trading the lower Time Frames is so demanding is because you must know what the Time Frames higher than those you are trading are telling you. There's a lot to keep up with! While it is extremely unlikely you'll be reaching a weekly '*'.382 Node when you're trading a five minute chart, it is *not* unlikely that an hourly or daily '*'.382 Node will affect your decisions. Be myopic at your own risk. Don't spend all you time looking for potholes while simultaneously ignoring the bridge that may be washed out just a little ahead.

Illustration of the above discussion would require copious examples and a huge amount of space. It is one of those areas best left to a classroom setting or repeated application by the individual trader.

CHAPTER 15

MORE MARKET EXAMPLES

□□□

A LONG TERM SOYBEAN MEAL TRADE:

Now let's consider a longer term trade in Soybean Meal. The approach outlined can help you uncover trades that are big winners over a period of months, not just over the next 15 minutes.

GENERAL DISCUSSION:

It's my habit to look over the weekly and monthly continuation charts of about 20 futures contracts, during the weekend. The markets are closed and I can get an objective, uncluttered viewpoint. I look primarily for Directional Indicators, major D-Levels, and of course I also look at the Trend indicators. In addition to determining the Trend, I look to see if there is *acceleration of price,* or *thrust*, through any of the DMAs I use, particularly the 25X5. If there is, I take notice. Strong thrust through the 25X5 on a continuation chart, particularly *after* a quiet period, often signifies a major move is on the way.

CONTEXT:

As I was paging my way through my continuation charts, I saw Chart 15-1 and became really excited. Here's what caught my attention.

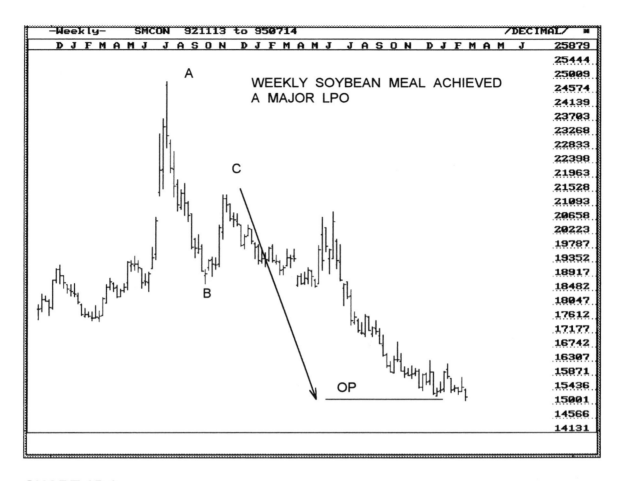

CHART 15-1

26 Apr 97 08:13:35   Updated: 00/00/0000   0.618 1.618
Point Value   Objective Points   File SMNWK2

    A = 24900   COP = 17598
    B = 18700   OP = 15230
    C = 21430   XOP = 11398

Copyright (c) 1996 CIS, Inc.

Consider the A, B, C expansion depicted in Chart 15-1 and in the accompanying
FibNodes™ printout. The OP move down was fulfilled and therefore indicated potential
support, almost exactly where it was when I first observed the chart. The potential
support, however, significant as it was, was *not* strong enough for me to take a position
even though the weekly MACD supported a long entry (Chart 15-3). Adequate context
for the trade was still lacking. There were no signs of up Movement or up thrust.

I hadn't traded meal since the early 80s. I had all but forgotten it was still traded. When I looked at this chart a little longer, however, I remembered that meal seasonally bottomed at about this time of year (March). With the additional consideration of seasonality, there *was* enough context for me to *want* to take a shot at the long side of this market.

TRADE IMPLEMENTATION:

Given the long term nature of the trade, the entry was easy. I would simply enter "at the market" since I didn't need to worry about getting a few cents off the price Where to place the stop was the problem! There was no sign of support, and the XOP was miles away, so I looked for an alternative approach. While I rarely use option strategies, this was a natural situation to employ one. Here's what I did. I bought the futures contracts "at the market" and bought puts against my position to lock in a maximum loss. I made these trades in the July contract so I would have plenty of time for things to go my way. For those of you who may be unfamiliar with option strategies, I'll elaborate. The cost of the put, whose strike price was near the current market price, was essentially the maximum loss I could sustain. That's because the puts gave me the right to sell contracts near the price at which I had bought the futures. If the price went up, I would profit on the futures and cover the puts. If the price went down I would profit on the put options and make up any loss on the long futures position. This type of strategy is called a "beach trade." You put it on and go to the beach!

Shortly after I put on the position, we had a Wash and Rinse and then acceleration of price through the 25X5. See Chart 15-2. The MACD/Stochastic combination shown in Chart 15-3 signaled a buy, so everything was in gear to the up side.

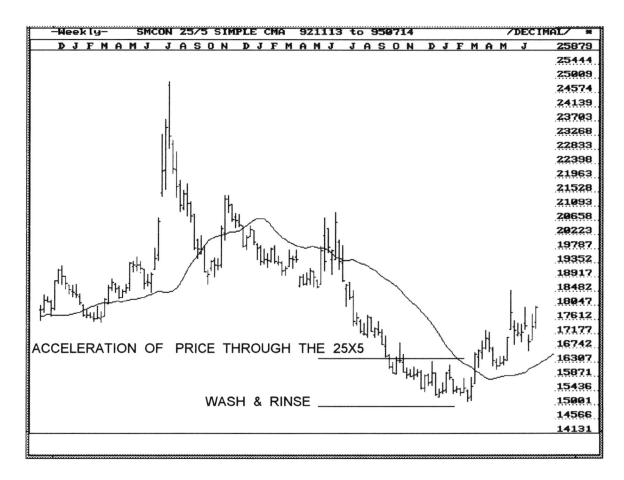

CHART 15-2

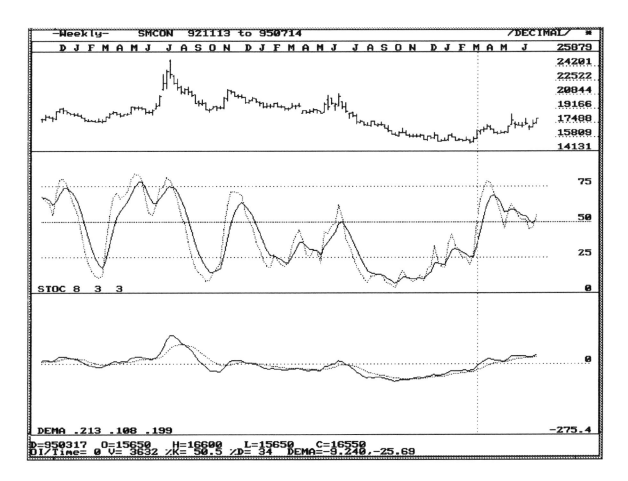

CHART 15-3

Even if you didn't use the more complex option strategy noted above and you knew nothing about the seasonals, you should have been ready to get on board. Why? Your attention would have been on the Trend indicators, due to the *fulfillment of the weekly OP*. Then, after the Trend signal, the up thrust, and the Wash and Rinse, you would have taken the *first* opportunity to enter.

Let's look at daily July meal Chart 15-4. We're at the same date as pictured in the weekly chart above. All we've done is reduce the Time Frame to a daily for a more detailed look.

The 3X3 is containing the up thrust, after the acceleration through the 25X5. For clarity, I've shown only a small portion of the 25X5. To segregate your thinking, I strongly recommend you never mix Trend indicators and Fibonacci work on the same chart. I'm breaking this rule in Chart 15-4 however, since the D-Level presentation is simple (only one Focus and Reaction Number) and I want to economize on space.

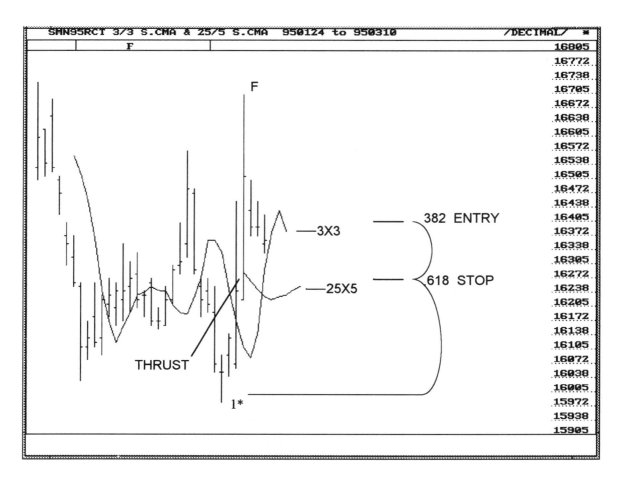

CHART 15-4

Where was this first opportunity to enter? Just look at the D-Levels after the acceleration (or thrust) through the 25x5.

Your entry would be just above the .382 Node. Your stop would be under the .618 Node. This is entry technique Minesweeper A, i.e. after the OP support and up thrust manifested on the weekly, you took the first dip to go long. The technique I used was a sophisticated Bonsai entry, using the context mentioned above, as well as that of seasonal support. My money stop was the cost of the put options.

Okay, we're long. Now where do we get out? The answer depends on your Time Frame. This could easily be a weekly-based trade, as it started out for me. In that case, you would go to a weekly chart, and generate an OP as the wave developed. You could use that OP level as your profit objective. If you were a daily-based player, you'd go to the daily chart to generate a daily-based OP. The second of these possibilities is pictured by the A, B, C move on the daily Chart 15-5 and shown numerically on the FibNodes™ printout below the chart .

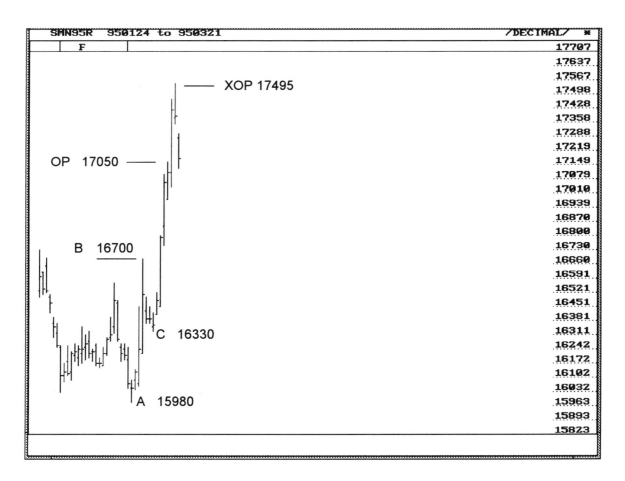

CHART 15-5

26 Apr 97 10:51:34   Updated: 00/00/0000   0.618 1.618
Point Value   Objective Points   File SMNDA1

A = 15980   COP = 16775
B = 16700   OP = 17050
C = 16330   XOP = 17495

Copyright (c) 1996 CIS, Inc.

As it turned out, the XOP was quickly achieved but there was nothing wrong with the nice OP profit. Shortly, I will go into detail on why the OP was my choice as an exit, rather than the XOP. That's right, we don't always get the high!

THE TRADE CONTINUES:

Okay, let's assume we're out at the OP. The D-Level™ series at this point is illustrated on Chart 15-6 showing an area of Confluence between 16796 and 16950.   The accompanying FibNodes printout represents the D-Levels existing at that point in time.

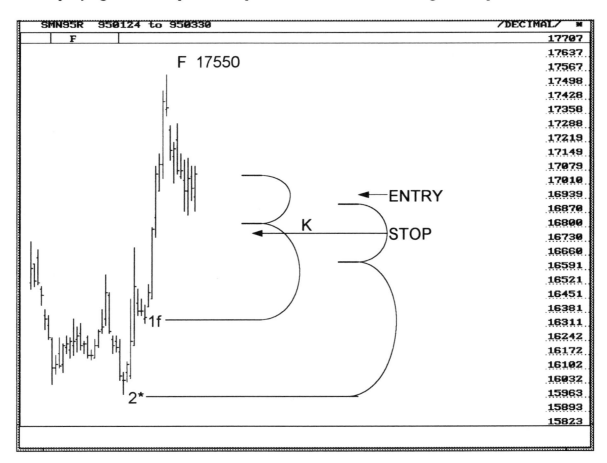

CHART 15-6

To reenter this move, it would be prudent to wait for a retracement to Confluence, although I'm sure Hyper Hank would have put an order in at the first retracement 17084.

```
26 Apr 97    11:43:39    Updated: 00/00/0000          0.382  0.618
Focus Number  File SMNDA2        Focus# (High for the swing)
Point Number  Support Fib Nodes    Point# (Enter highest reaction low first)
-------------------------------------------------------------------------
17550    1
      ===>  17084
      ===>  16796
16330

-----------------------------------
17550    2
      ===>  16950  *
      ===>  16580  *
15980

-----------------------------------
```

If we assume our entry was just above the top end of the Confluence area and our stop below the bottom end, we would have been filled and our stop would have remained untouched. Alternatively, it would have been acceptable to put an initial stop under the Primary '*' .618 Node at 16580 and handle any break of Confluence in the manner described under "Advanced Comments" in the bond Double RePo trade discussed in CHAPTER 12. Now we're in and ready for the next up move.

Below are *two* Fibnode expansion series: the Fibnode expansion originating from the low at 15980 and the second series originating from the low at 16330. Both are valid. See Chart 15-7A.

So, what do we do with six profit objectives ?

To simplify the discussion, let's consider the second expansion which is producing lower profit objectives than the first.

```
26 Apr 97  15:42:32          Updated: 00/00/0000    0.618  1.618
Point Value    Objective Points    File SMNDA03
-------------------------------------------------------------------------

    A = 15980        COP = 17830
    B = 17550         OP = 18430
    C = 16860        XOP = 19400
```

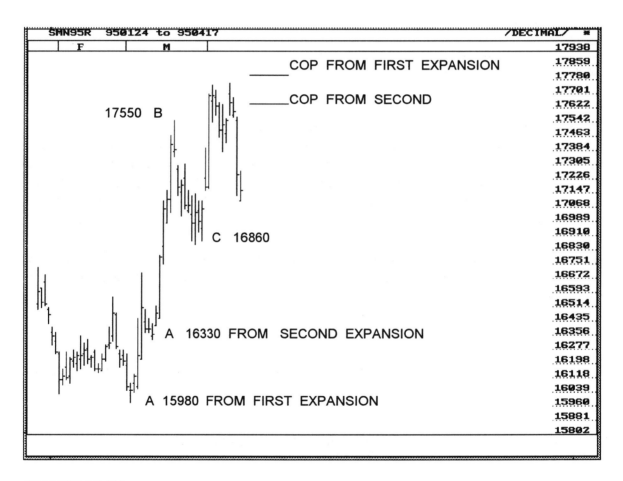

26 Apr 97 15:43:12   Updated: 00/00/0000   0.618 1.618
Point Value   Objective Points   File SMNDA5

     A = 16330   COP = 17614
     B = 17550   OP = 18080
     C = 16860   XOP = 18834

Copyright (c) 1996 CIS, Inc.

CHART 15-7A

We have three profit objectives to choose from. In hindsight, it's obvious the COP would
have been the right choice. It would have been my choice *before the fact* as well. Here's
why. Remember I was willing to take the OP profit on the first move up from the 15980
level. For the same reason, I would be conservative and take the COP out of this move.
When a market breaks down for weeks and months as Soybean Meal did, it usually has
some difficulty initially coming out of the hole, as overhead supply in cash markets comes
to bear.

Distant profit objectives are initially difficult to achieve and typically there's a '*'.618 retracement lying somewhere along the line, after the first move up. Later when more traders and commercials realize the true structure of the move, we can be more aggressive in what we are willing to take out of the market and have more confidence in doing so.

Now, let's consider the first expansion indicating COP resistance somewhat higher at 17830. For the same reasons as those just cited, I would take the more conservative COP of the second expansion. There is another reason, as well. Take a look at Chart 15-7B.

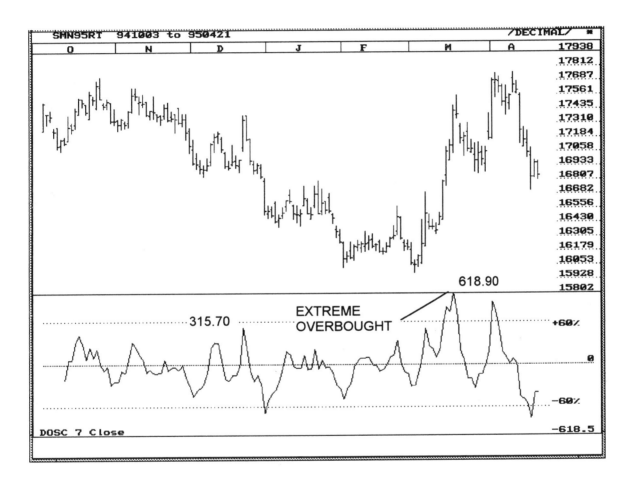

CHART 15-7B

In CHAPTER 7, we discussed taking close in profit objectives, when the market was Overbought. The extreme overbought area represented by the 618.90 Detrended Oscillator value certainly qualified. That's precisely the situation the market was in as the COP profit objective was approached by price!

THE TRADE CONTINUES:

After the COP was achieved, and we booked another profit, we finally got to the '*' .618 Primary Node, in the form of a retracement down to 16630. Here's where the puts come off, and long positions can again be established for the OP move to 18410. This OP move is discernible on both daily and weekly charts. Chart 15-8 and the accompanying FibNodes printout shows how you arrive at the Fibnode series, while Chart 15-9 shows the retracement back to 16630, as well as the expansion to 18410.

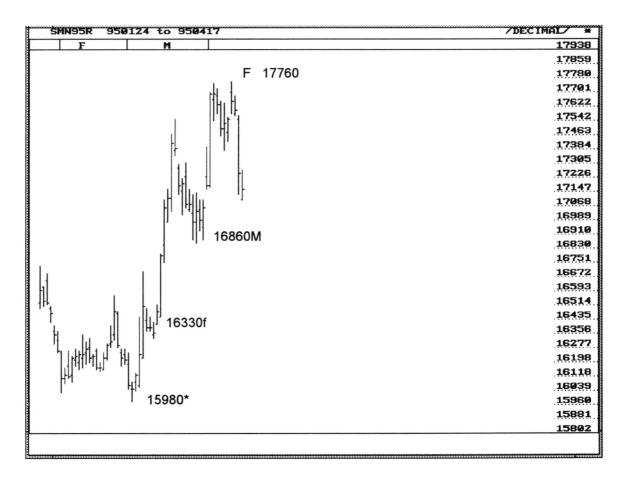

CHART 15-8

Focus Number   File SMNDA4   Focus# (High for the swing)
Point Number   Support Fib Nodes   Point# (Enter highest reaction low first)

17760   1
        ===> 17416 M
        ===> 17204 M
16860

17760   2
        ===> 17214 f
        ===> 16876 f
16330

17760   3
        ===> 17080 *
        ===> 16660 *
15980

Copyright (c) 1996 CIS, Inc.

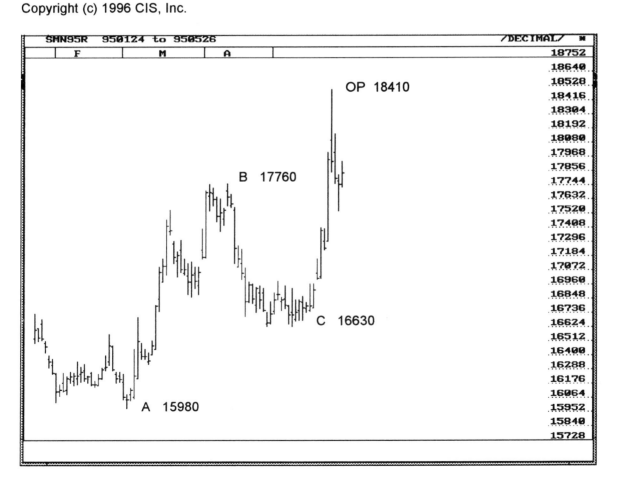

CHART 15-9

IMPORTANT POINTS TO NOTE:

Nowhere in this trade am I interested in playing the short side, even though my techniques indicated pretty good places to take a shot at the short side. Why? The percentages aren't there. Remember, we're coming out of the hole, after a major weekly OP, a seasonal low, and thrust through the DMAs. Why fight all that!

ADVANCED COMMENTS:

Trading in hindsight is always easy. *Trading decisions based on what you know at the present time are what count.* Reconsider the first OP profit objective at 17050, in light of what was shown on Chart 15-7B. The maximum Overbought oscillator up to that point was 315.70. The OP (price) was in that vicinity of overbought when I chose to take a profit. In hindsight, the market went to roughly double that value, *but I didn't know that it would,* even though the context for the trade suggested that it might. Once the previous oscillator value was achieved, I had an additional reason for getting long this market by using "Special Applications of the Detrended Oscillator" Strategy 5, in CHAPTER 7.

Later in this example, I was willing to take the COP out at 17614. I was looking at the *new overbought extreme of 618.90,* not the old value of 315.70!

Note that my second trade entry just above the top end of Confluence at 16950 was very close to my initial exit at the 170 area. *In hindsight*, I might as well have stayed in! We don't have the advantage of hindsight, however, when we trade.

The reentry at 16950 was a much safer trade than staying with the position at the 170 area because of all the reasons cited above.

In trading based on technical analysis, you are concerned with percentages, with the *likely Movement of price*, not the value of the price itself. Buying a dip at 500 after a consolidation may be better than buying a peak at 400 after thrust, *because it's safer!* Don't make it hard on yourself. Play the percentages and whistle all the way to the bank.

WAS A MISTAKE MADE?

Even though this was planned as a long term trade that I expected to last for months, I traded it on a daily basis. I took advantage of daily swings. Was this a Mistake? No, I don't think so. This market was easy to read and ready "for the pickin's." While I was actively trading other intraday markets, there was nothing so captivating that I couldn't take the time to focus on my old friend, meal, for a few minutes each day.

MORE EASY PICKIN'S:

Note: Do you see the Bread & Butter trade represented by the initial thrust up on Chart 15-7A (3X3 not shown)?

Note: The move down from the COP was fast and furious. It looked as though a Double RePo (sell) was apparent. The time between the first and second penetration bars, however, was just too wide to qualify and a Look-Alike wasn't enough to motivate me to sell this market, in light of all the positive factors.

A SHORT TERM S&P TRADE:

I particularly like this S&P example. Maybe it's because of the money I made trading it, or maybe it's because it's such an illustrative example. One thing's for sure; it's indicative of the way I've approached trading S&P since 1985. It's the way I like to interact with volatile markets.

GENERAL DISCUSSION:

First we will consider the *context* for the trade. Then, we'll examine how trade psychology and market mechanics play their role. We will go through a step by step process, showing each aspect of the Trading Plan and how it was implemented for this low risk, high profit, hit-and-run experience.

TREND:

From the following daily Chart 15-10, you can see we've been above the 25X5 for months, above the 7X5 for over a week, and it's obvious from the price action that we have been and are currently above the 3X3 as well. Therefore, the daily Trends are all up. Furthermore, based on the value of the various DMAs, they are likely to remain up "tomorrow," unless we get a very severe break in price.

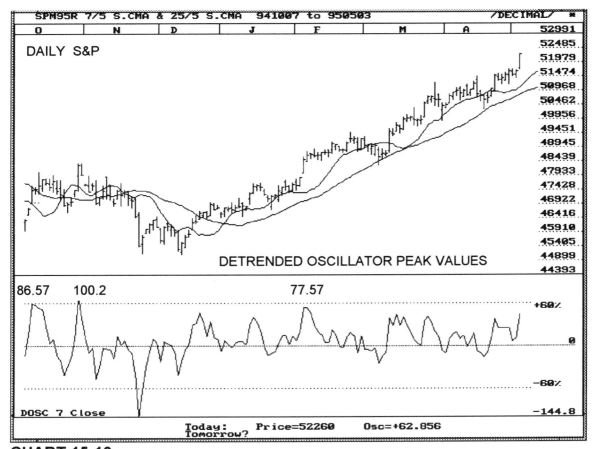

CHART 15-10

Tomorrow

7/5   MA = 51174, Q (basis OSC= 7) = -43.40851
25/5 MA = 50858, Q (basis OSC = 7) = -70.42389
 3/3 MA = 51578, Q (basis OSC = 7) = -8.71582

Price of 52849 would produce Q+MAX (OSC = 7) = +100.2139
Price of 49990 would produce Q+MAX (OSC = 7)= -144.8575

(c) Copyright CIS Inc. / Microforce

Before we go to the next step, I want to digress for a moment to explain the numbered printouts associated with this chart[1]. To begin with, the chart was printed as of "last night's close," the day *before* the trade took place. The group of numbers below the chart shows the Oscillator Predictor™ values at various prices. We are most interested in the maximum Overbought and maximum Oversold levels.

---

[1] The chart produced from the "TIMESAVER", CIS TRADING PACKAGE

A price of 52849 would produce an "historically" maximum Overbought level of +100.2139. The abbreviation used for this is Q+max . "Historically" refers to the last six months. The maximum Oversold level would be achieved at a price of 49990, which is a Detrended Oscillator value of -144.8575. If you don't have the Oscillator Predictor, you can use alternative methods described in CHAPTER 7.

OVERBOUGHT AND OVERSOLD ANALYSIS:

If you take the three peaks of the Detrended Oscillator and average them together (86.57 + 100.20 +77.57, divided by three), you get 88.11. I typically eyeball the values and estimate an average. Let's use 90. The Oscillator Predictor point shown in Appendix Q for this value produces a price of 52730. Conclusion: *this contract is going to be at a maximum stretching point tomorrow, if it achieves a price anywhere between 52730 and 52849.* As of the close (last night), we were at about 62% of maximum Overbought, so this number was high enough to bear watching.

DIRECTION:

I don't see any obvious Directional Indicators in play, with the possible exception of Stretch. What I mean is that when I looked at this chart on the night before the trade took place, I concluded that there were no Double RePos, RR tracks, Head and Shoulders, Failures etc., but that it was possible we *could* end up with a Stretch before tomorrow's action was over, since we were approaching Overbought. I hadn't calculated any expansions to the upside, or resistance Fibnodes, so I wasn't sure if Stretch was a real possibility. Nonetheless, this was duly noted and attached to my S&P clipboard.

TRADE IMPLEMENTATION:

The next morning the market opened strongly by gapping up, then it made a nice .382 retracement back to the 52230 level and continued on to the upside. You can see this by skipping ahead to the five minute Chart 15-16. Given the context, the only possibility *early in the day* was to play this market to the upside.

Look at the MACD/Stochastic on the daily, hourly, and half-hourly charts. They all pointed up until at least 13:00.

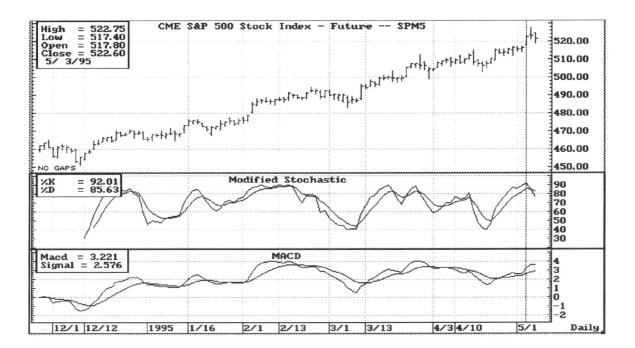

CHART 15-11

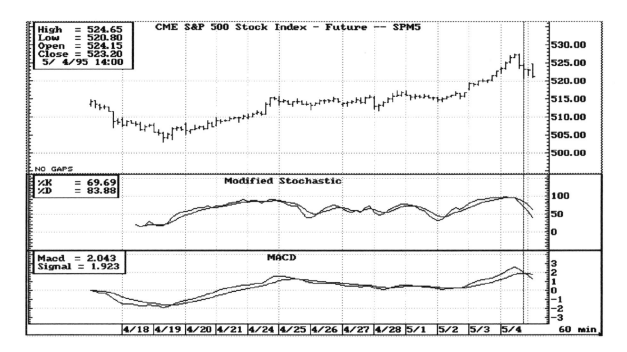

CHART 15-12

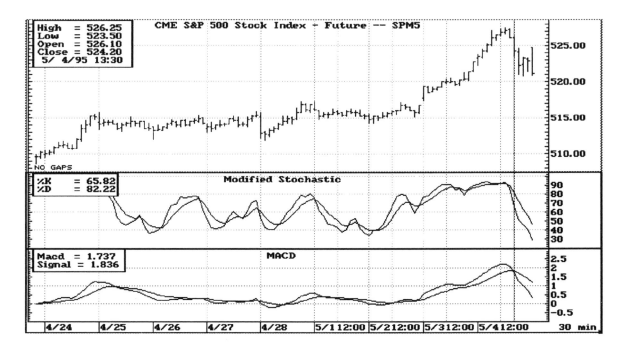

CHART 15-13

The problem with taking a long position, however, was that price was reaching very high Overbought levels. Therefore the risk that it would not continue higher was significant.

Also, any profit objective on a long entry would have to be close. The bottom line was that the risk/reward ratio just wasn't positive enough for me to get involved. Given these facts, my focus was on the bond market, since the day trading opportunities there were much more favorable.

After a good day trading the bond market, I took another look at the S&P. It was about 14:15 to 14:30. The high stood at 52745, *clearly in the highly overbought range*. I noticed that by 14:00 both the MACD and Stochastic on the half-hour chart were in a sell mode and the Stochastic on the hourly was also in a sell. The hourly MACD was approaching a sell. All these facts taken together told me we had an excellent chance for a rout to the down side before the day ended, catching all the (long side) Johnny-Come-Latelies unaware. Given the look of the Trend indicators at about 14:15, if it were going to happen, it would take place soon.

Okay, everything I talked about so far is the *context* for the trade. This is the kind of reasoning I go through on each and every trade. It is how I evaluate my risk/reward. It is how I determine whether or not I want to play a given market. The bottom line was that I had strong downtrends forming on intraday charts after reaching a significant Overbought level.

TRADE PSYCHOLOGY:

Before getting into the details of entry and exit, let's evaluate the psychology of the market at this point in time.

Long side players have been rewarded day after day. Many have doubled and tripled up positions attempting to make millions out of a few thousand. Few participants are using profit objectives: after all, the books tell you to let your profits run, to let the market take you out. Some players are so confident, they've thrown in a stop and gone to the golf course. Others who have gotten out at a nice profit are so frustrated at the profits they've "lost" (greed), that they have gotten back in with larger positions to "make up for" their "error" in getting out too soon.

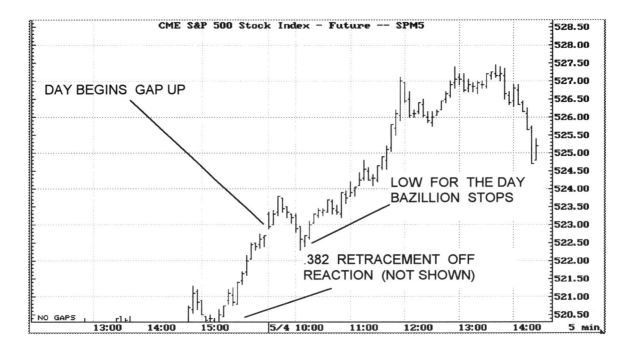

CHART 15-14

MARKET MECHANICS:

If you know anything about market mechanics, you'll know that by midday there are a *bazillion stops* under the low of the day at 52230. You also should know that's the area where the locals will be gunning for stops if we get any kind of Movement south. To make that last statement more understandable, here's what happens. They (the locals) sell the market mercilessly on the way down and when the sell stops are hit, they buy them, *after, and during the panic*, thereby getting flat and achieving a nice profit. Where do you think all the Mercedes, Jags, and BMWs come from in the MERC'S parking garage? It's their job to facilitate trade, provide a liquid market, and buy Mercedes, Jags and BMWs. Traders off the floor don't realize it's our job to *join in with them*. So here's how we do it.

CHART 15-15

Let's assume you've looked at this trade late in the day, as I did. Notice, you had a little Wash and Rinse at Point 3 '*', another encouraging sign to the down side. By the time it was 14:30, a strong down wave had developed. FibNodes™ Focus and Reaction Numbers are labeled on Chart 15-15. Notice the Confluence area 52467-52480. Selling a reaction back was as easy as it gets.

```
25 Apr 97     09:19:03    Updated: 00/00/0000          0.382  0.618
Focus Number   File SPM051          Focus# (Low for the swing)
Point Number   Resistance Fib Nodes   Point# (Enter lowest reaction high first)
---------------------------------------------------------------------------
52350     1
     ===>  52423   T
     ===>  52467   T
52540
------------------------------------
52350     2
     ===>  52480   f
     ===>  52560   f
52690
------------------------------------
52350     3
     ===>  52501   *
     ===>  52594   *
52745
------------------------------------
Copyright (c) 1996 CIS, Inc.
```

Selling "K" was an ideal and "safe" area to take a shot at the down side. You could select any one of a number of "Bushes" stops, but at a minimum I'd hide my stop above the '*'.382 Node at 52501. There would also be nothing wrong with selling the first Node back (box 1) at 52423, instead of the Confluence area. It depends upon how aggressive you are. I chose to sell at the lower end of Confluence at 52465 and also to sell stop the low at the Focus Number F, 52350 (chart Point B). While I normally don't initiate orders on stops, I chose to do so in this case, because I knew there would be some locals and intraday players closing short positions (buying) at the old lows 52350, 52320, and 52230. I also knew my broker commanded enough respect in the pit to get me filled reasonably well, if any of these buy orders manifested.

While my minimum expectation for the trade was to take out the low of the day at 52230, I quickly calculated the expansion down, which yielded an OP at 52070. This expansion is detailed in Chart 15-16 and shown in the accompanying FibNodes printout.

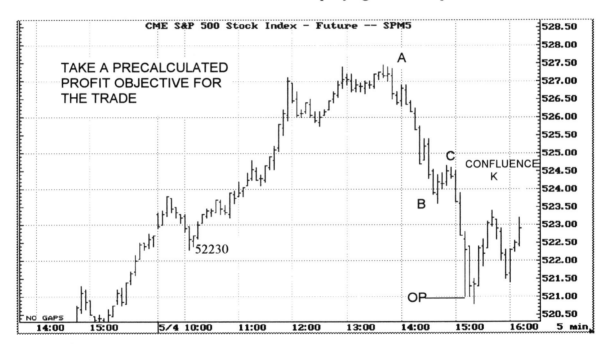

CHART 15-16

25 Apr 97 12:24:23   Updated: 00/00/0000   0.618 1.618
Point Value   Objective Points   File SPM052

    A = 52745   COP = 52221
    B = 52350   OP = 52070
    C = 52465   XOP = 51826

Copyright (c) 1996 CIS, Inc.

Why did I choose the OP? Well, the COP was just below the low of the day and that's where all the sell stops were. If we hit that point there was no way the COP would hold. The XOP at 51826 was a mile away and I reasoned that before reaching it there was likely to be a throw back rally. The OP was most likely to be hit by the simple process of elimination.

Let's look at the way the trade worked out. The Confluence area, where my initial sell order was placed, was hit to the tick at 52465. What's more interesting, however, is the second entry at 52350.

SPM5 – CME S&P 500 Stock Index – Future							Time & Sales
5/ 4/95							
15:05 523.75	523.70	523.65	523.60	523.65	523.70	523.60	523.65
523.70	523.75	523.80					
15:06 523.75	523.80	523.75a	523.70	523.65	523.60	523.65	523.70
523.75	523.70	523.75	523.80	523.85	523.90	523.85	523.80
15:07 523.75	523.70	523.65	523.60	523.55	523.50	523.45	523.40
523.35							
15:08 523.40	523.35	523.30	523.25	523.20	523.15	523.10	523.15
523.20	523.30	523.25	523.20				
15:09 523.15	523.10	523.20	523.15	523.10	523.05	523.00	522.95
522.90	522.85	522.80					
15:10 522.75	522.80	522.75	522.70	522.65a	522.60	522.50	522.45
522.40	522.50	522.55	522.60	522.65	522.70		
15:11 522.60	522.70	522.75	522.80	522.90	523.00	523.05	523.10
523.00	522.90	522.80	522.75	522.70	522.80	522.90	523.00
523.10							
15:12 523.00	521.90	521.70	521.60	521.50	521.40	521.30	
15:13 ▶ 521.20	521.10	521.00	521.10	521.20	521.30	521.50	521.60
521.70	521.80	521.90					
15:14 522.00	521.70	521.50	521.60	521.70	521.80	521.90	
15:15 522.00	521.95	521.90	522.00	521.50	522.00	522.10	522.20
522.30	522.50						
15:16 522.40	522.00	522.20	522.30	522.50	522.40	522.20	522.10 →

CHART 15-17

ADVANCED COMMENTS:

Notice the time and sales[2] Chart 15-17 and the *way* the market fell. From 52390 at 15:06 to 52335 at 15:07 there wasn't an up tick. Exchange rules say you aren't *due* a fill until you get an up tick (on a sell). That happened at 15:08 at 52340. That's where my second order was filled, at 35. Now look at the action several minutes later. Note 15:12. *We go from 52300 to 52190 in one tick!* We continue with the glide ratio of an anvil to the 52100 area. It's no accident support finally came in 30 points above the OP! Let's take another look at the five minute Chart 15-16. You'll see the test of the 52100 low was at

---

[2] Time and sales in Chart 15-17 is what my data service provider, my computer, and my software, displayed. It was what I was using to trade. Appendix H shows time and sales, gathered after the heat of battle, from a different "more reliable" source. It's interesting to note the difference.

52075, just one tick above the OP! What happened was that the 52100 test, down to 52075, was just enough to wash out the stops below 52100 and reach solid support at the OP. So how did I handle it?

When I saw the tick from 300 to 190, it was Pampers® time! Hundred point ticks weren't all that common at the time. I called the floor to see if the tick was correct. Their voices were hoarse and shaky. The order clerks told me there was a trade or two between those two numbers, but that a panic was occurring. "We're not 'arbing' orders, it's so crazy," was what the clerk said. The noise level was extreme. The clerks were concerned with errors, and rightfully so. Before I got off the phone we were at 52100 and since that was just a few ticks above my closing profit objective, and since there was a panic going on, I "canceled replaced" and closed at market. Many of my students and my trading friends would have been delighted to see a 110 point tick in their favor. For me it was the ticket out. That old saying about Fools and Angels ought to hang on your trading room wall.

When I know I'm trading really well and in tune with the market, this is the way things go. Hit the market on an extreme in a throw back rally, never suffer pain, and take profit in just a few bars. If you think that this sounds just too good to be true, use these strategies for a while and see for yourself just how close you can get to the "perfect trade." It might surprise you.

A QUESTION:

This sounds fantastic. Do you always trade like this?

Sure... I *always* sell highs, I *always* buy lows, my stops are *never* hit, I own Chicago and New York. My bid is in for Singapore and I'm looking for a retracement!

EPILOGUE

For those of you who are wondering about what became of Hank, Dan, and Carl, I'll fill you in on what I know.

Hank's still hard at it. He's a broker now and has an advisory service. His clients last longer than most who trade futures, but they sure seem to trade a lot.

Carl gave up futures trading and runs a nursery specializing in petunias. He managed to make some money trading, but it was too stressful and considering the amount of work and worry it entailed it just wasn't worthwhile. His strawberry petunia display won a blue ribbon at the county fair.

What about Dan?

We'll, he kept slipping through the rain drops and made a lot of money in Futures. For those jealous ones among you, take heart. After one of his frequent trips to Bangkok, his wife filed for divorce and the settlement was a doozie. Problem was however, she's been unable to collect. Something about offshore banks.... *He sure is a tough guy to tag.* He's still trading but nobody knows for sure where he is.

APPENDIX A

CALCULATIONS AND CHART LOCATION FOR THE 3X3 DISPLACED MOVING AVERAGE

DEFINITION:

The 3X3 Displaced Moving Average is a number which is derived by computing the three day simple Moving Average of the close, and displacing it forward in time three days. A series of these numbers, connected together, form a Trend Indicator, as well as a tool for determining certain directional signals.

EXAMPLE:

If we start with 95/06/12, we can calculate the 3X3 for the Sept. S&P on 95/06/19.

DATE - 95	CLOSING PRICE	3X3 DMA
6 - 12	536.80	
6 - 13	540.80	
6 - 14	540.60	
6 - 15	542.80	
6 - 16	543.90	
6 - 19	549.65	539.40
6 - 20	549.30	541.40
6 - 21	548.95	542.43
6 - 22		545.45
6 - 23		547.62
6 - 26		549.30

We simply add the closing price on 6/12 to the closing price on 6/13. To that sum we add the closing price on 6/14. We then divide that total by three. This is what is commonly known as the simple 3 day Moving Average of the close. Conventional wisdom would place that number on the chart on 95/06/14. We "displace" it forward 3 (market) days. Therefore the number calculated represents the 3X3 DMA for 95/06/19.

APPENDIX B

ㅁㅁㅁ

RUNNING FibNodes™ AND TradeStation® AT THE SAME TIME

Perform the following steps to run *FibNodes* and *TradeStation* at the same time, *after* the programs have been installed using each program's installation instructions:

Windows® 3.1 Users:

1. After *TradeStation* is running, hold down the **CTRL** key while pressing the **ESC** key on your keyboard to display the Windows task list.

2. Select **Program Manager** from the list by clicking on it and click the **Switch To** button to display the Windows Program Manager.

3. Click the **Window** menu.

4. Select **Main** from the list that is displayed to open the **Main** program group.

5. Double-click the **MS-DOS Prompt** icon to access DOS.

6. Follow the normal procedures to load the *FibNodes* program.

7. After *FibNodes* is loaded, hold down the **ALT** key while pressing **Enter** on your keyboard to open a DOS shell.

8. To view *TradeStation* and the DOS window simultaneously, arrange the windows horizontally on your screen so that both applications are visible.

9. If you need to shut your computer down at any time, go back into DOS and exit the *FibNodes* program, and then type **exit** at the DOS prompt.

Windows® 95 Users:

1. After *TradeStation* is running, hold down the **CTRL** key while pressing the **ESC** key on your keyboard to display the Windows **Start Menu**.

2. Select **MS-DOS Prompt** from the menu to access DOS.

3. Follow the normal procedures to load the *FibNodes* program.

4. After *FibNodes* is loaded, arrange the DOS window so that both *TradeStation* and *FibNodes* are visible on your screen.

5. If you need to shut your computer down at any time, go back into the DOS window and exit the *FibNodes* program, and then type **Exit** at the DOS prompt.

APPENDIX C

FibNodes™ setup for *ASPEN GRAPHICS™ USERS*
Windows® 3.1

These instructions are designed to help users of FibNodes and ASPEN GRAPHICS run the two programs side-by-side on a Windows 3.1 operating system.

1) Install FibNodes and ASPEN GRAPHICS using each program's installation instructions.
2) Bring up Aspen as you would normally (i.e. as a full-screen application).
3) Press the alt+tab keys. to get back to Program Manager.
4) Find the Group Icon that contains the MS-DOS icon (usually **Main**). Highlight this icon, press the enter key, and then highlight the MS-DOS icon and press the enter key.
5) At the DOS prompt, type: **cd\fib** and press the enter key.
6) At the C:\FIB prompt, type: **fibnodes** press the enter key.
7) Press the enter key four (4) times to move past the introductory screens.
8) Press alt+enter keys to make the FibNodes program run in a small, inset window. Adjust the window size using the resize icon in the upper-right corner of this window.
9) Switch back to ASPEN GRAPHICS using the alt+tab keys.
10) Adjust the size of the ASPEN window by clicking on the resize icon. Use the cursor to further adjust the size of the window by holding down the left mouse button and dragging the borders until both windows can be viewed on the screen side-by-side.
11) Switch between programs by using the alt+tab keys, or click the right mouse button in the window you want to use.

FibNodes™ SETUP FOR *ASPEN GRAPHICS™ USERS*
Windows® 95

These instructions are designed to help users of FibNodes and ASPEN GRAPHICS™
run the two programs side-by-side on a WIN95 operating system.

TO OVERLAY FibNodes™ ON ASPEN GRAPHICS™:

1) Install FibNodes and ASPEN GRAPHICS using each program's installation
   instructions.
2) Bring up Aspen as you would normally (i.e. as a full-screen application).
3) Click-left on "**Start**" on the Taskbar, usually found in the lower-left corner of
   the screen.
4) Select **Programs** and from the selection of programs select **MS-DOS Prompt.**

If DOS appears to take over your entire screen:

Press alt+enter to make the MS-DOS Prompt program run in a small, inset
window. Adjust the window size in the upper-left corner of this window.

5) Type: **cd\fib** and press the enter key.
6) At the C:\FIB prompt, type: **fibnodes** and press the enter key.
7) Press the enter key four (4) times to move past the introductory screens.
8) Switch between FibNodes and ASPEN GRAPHICS using the alt+tab keys.

TO VIEW FibNodes™ AND ASPEN GRAPHICS™ SIDE-BY-SIDE ("TILED" ON YOUR SCREEN):

1) Bring up ASPEN GRAPHICS™ as a full screen application and bring up
   FibNodes in an MS-DOS window (Steps 1 through 8 above).
2) Click either mouse button anywhere within the ASPEN GRAPHICS™ window.
3) Click on the resize icon in the upper-right corner of the screen.
4) Once the ASPEN GRAPHICS™ window has been resized, click on the resize
   icon in the MS-DOS/FibNodes window.
5) Move and resize windows as necessary until both windows are visible at the
   same time.
6) To switch between windows, use the at keys or click the mouse in the
   window you want to use.

APPENDIX D

ccccccccccccccccccccccccccccccccccccccccccccccccccccccccccccccccccccc

TradeStation® INPUTS TO SIMULATE STUDIES IN *DiNapoli Levels™.*

MOVING AVERAGE CONVERGENCE DIVERGENCE (MACD)[1]

To use smoothing factor inputs of .213, .108, and .199 where the number of "periods" is required, use the preprogrammed MACD (which is calculated by an exponential MA formula) with the following inputs. The formula for MACD is in Appendix E.

FAST MA	*8.3896*
SLOW MA	*17.5185*
MACMA	*9.0503*

DISPLACED MOVING AVERAGES

3 Period Simple MA of close Displaced Forward 3 Periods
7 Period Simple MA of close Displaced Forward 5 Periods
25 Period Simple MA of close Displaced Forward 5 Periods
Period = Day, Week, or Month

DETRENDED OSCILLATOR

Close - N Period Simple Moving Average
Period = Day, Week, or Month
N = 7 or 3

---

[1]Gerald Appel, *The Moving Average Convergence Divergence Trading Method* (New York: Signalert Corporation).

PREFERRED STOCHASTIC

PREFERRED SLOW %K USER FUNCTION

* *

Study	*: PreferredSlowK*
Title	*: Slow %K based on Modified Moving Average*
Type	*: User function*
Note	*: FastK is preprogrammed*

* *

Input: SlowKLen(Numeric), FastKLen(Numeric);
PreferredSlowK = PreferredSlowK[1] + ((1/SlowKLen)(FastK(FastKLen) -*
PreferredSlowK[1]));

PREFERRED SLOW %D USER FUNCTION

* *

Study	*: PreferredSlowD*
Title	*: Slow %D based on Modified Moving Average*
Type	*: User function*
Note	*: FastK is preprogrammed*

* *

Input: FastKLen(Numeric), SlowKLen(Numeric), SlowDLen(Numeric);
PreferredSlowD = PreferredSlowD[1] + ((1/SlowDLen)(PreferredSlowK(SlowKLen,*
FastKLen) - PreferredSlowD[1]));

PREFERRED STOCHASTIC INDICATOR

* *

Study : *StochasticPreferred*
Title : *Stochastic based on modified Moving Average*
Type : *Indicator*

* *

Input: FastKLen(8), SlowKLen(3), SlowDLen(3);
Plot1(PreferredSlowK(SlowKLen, FastKLen), "%K");
Plot2(PreferredSlowD(FastKLen, SlowKLen, SlowDLen), "%D");
If CheckAlert then
begin
If Plot1 crosses above Plot2 or
Plot1 crosses below Plot2 then
Alert = TRUE;
end;

PREPROGRAMMED TRADESTATION® STOCHASTIC INDICATORS AND FUNCTIONS

TradeStation® has four Stochastic *indicators* included in version 4.0 build 18. They are:

- Stochastic Fast
- Stochastic Slow
- Stochastic Fast Custom
- Stochastic Slow Custom

These indicators are plotted from the calculations of several *functions*, which are:

- FastK
- FastD
- SlowK
- SlowD
- FastKCustom
- FastDCustom
- SlowKCustom
- SlowDCustom

Suffice it to say, that aside from the formula for *FastK* (RAW), none of these Stochastic functions, and thus their associated indicators, are the published definition of George Lane's *Stochastic Process*, but are modifications of the original formula. Make sure that you check the listings for these functions using the TradeStaton® PowerEditor™, and know what you are using before making trading decisions that rely upon them.

APPENDIX E

FORMULAS AND STUDIES:

LANE FAST STOCHASTIC

According to John Murphy[2]:

%K = 100 [(C-L$_n$)/(H$_n$-L$_n$)]

%D = 100(H$_m$/L$_m$)

where:
C is the latest close
L$_n$ is the lowest low for the last n days
H$_n$ is the highest high for the last n days
H$_m$ is the m-day sum of (C-L$_n$)
L$_m$ is the m-day sum of (H$_n$-L$_n$)

FAST STOCHASTIC

According to Perry Kaufman[3]:

% K same as shown above

%D = (% K$_t$ + % K$_{t-1}$+ %K$_{t-2}$) / 3 (this a simple ma smoothing)

where:
%K$_t$ is %K for the most recent period

---

[2] John Murphy, *Technical Analysis of the Futures Market* (New York: New York Institute of Finance, 1986).
[3] Perry Kaufman, *The New Commodity Trading Systems and Methods* (New York: John Wiley & Sons, 1987).

FAST STOCHASTIC,
USING THE MODIFIED MOVING AVERAGE (MAV) FOR SMOOTHING
(Kaufman *New Commodity Trading*)

$$MAV_t = MAV_{t-1} + (P_t - MAV_{t-1})/n$$

where:
MAV_t *is the current modified Moving Average value*
MAV_{t-1} *is the previous modified Moving Average value*
P_t *is the current price*
n is the number of "periods"
The starting point is calculated identically to that of a simple Moving Average.

FAST STOCHASTICS:

$$\%K = 100\ [(C-L_n)/(H_n-L_n)]$$

where:
C is the latest close
L_n *is the lowest low for the last n days*
H_n *is the highest high for the last n days*

$\%D = 3$ *period modified Moving Average of* $\%K$

THE PREFERRED (SLOW) STOCHASTIC

$\%K = \%D$ *(from the fast stochastic above)*

$\%D = 3$ *period MAV of* $\%K$

MOVING AVERAGE CONVERGENCE DIVERGENCE (MACD) FORMULA
(Appel *The Moving Average*)

$EMA1_t = EMA1_{t-1} + SF1(P_t - EMA1_{t-1})$

$EMA2_t = EMA2_{t-1} + SF2(P_t - EMA2_{t-1})$

$MACD = EMA1 - EMA2$

$SIGNAL\ LINE = MACD_{t-1} + SLSF(MACDt - MACDt_{-1})$

where:

$EMA1_t$ & $EMA2_t$ are the two exponential Moving Averages current values.

$EMA1_{t-1}$ & $EMA2_{t-1}$ are the previous values for these EMA

$SF1$ & $SF2$ are the smoothing factors for $EMA1_t$ & $EMA2_t$

$MACD_t$ is the current MACD value

$MACD_{t-1}$ is the previous MACD value

$SLSF$ is the signal line smoothing factor

P_t is the current price

EXPONENTIAL MOVING AVERAGES
(Kaufman *New Commodity Trading*) [4]

$EMA_t = EMA_{t-1} + SF(P_t - EMA_{t-1})$

where:

EMA_t is the current exponential Moving Average value

EMA_{t-1} is the previous EMA value

SF is the smoothing factor

P_t is the current price

"approximate" smoothing factor $= 2/(n + 1)$

where:

n is the equivalent number of "periods" in a simple Moving Average

---

[4] J.K. Hutson, "Filter Price Data: Moving Averages vs. Exponential Moving Averages," *Technical Analysis of Stocks & Commodities* magazine, May/June 1984.

APPENDIX F
 oooooooooooooooooooooooooooooooooooooooooooooooooooooooooooooooo

ABOUT FibNodes™:

The FibNodes software has been designed to efficiently implement the strategies taught in DiNapoli Levels. It is exactly the same software I use for my own trading. There are no changes, no differences, and no secrets. It is used extensively in this country and abroad by savvy individual traders and money managers alike, who trade everything from Kuala Lumpur rubber to the S&P.

HOW FibNodes™ WORKS:

FibNodes is data *independent*. It *does not* use a data-base or on-line service to develop its files. *You* input the pertinent points. The *program* develops, stores, calculates and presents the support and resistance levels as described in DiNapoli Levels. If the market makes a new high or low, or if a new Reaction Number is called for, you add *only this number*; the program uses the data already in the file to recalculate all pertinent points. The presentation includes Lineage, however Confluence is *not* automatically determined, since Time Frame and volatility in a given market can account for widely disparate Confluence locations.

OPERATING SYSTEM:

Currently FibNodes is a DOS program. A Windows/NT version is being considered. There is little advantage to a Windows version, except for marketing purposes, since the DOS program is fully functional (in a DOS window) in WIN3.1, WFWG3.11, or WIN95. We supply documentation on how to display FibNodes in this manner. Of the current users, about half use the program in a separate computer, while the other half use a DOS window in the same computer they run their graphics software. I prefer to use a separate computer. It's really up to you. If you display FibNodes in Windows with your graphics software, I recommend a 17 to 25 inch monitor so you don't have to squint to see both your charting software and FibNodes at the same time. Another alternative is to run multiple monitors, on the same computer, and display FibNodes on one of them.

FibNodes™ REQUIREMENTS:

Since the FibNodes software is a sophisticated and narrowly defined calculator, its requirements are minimal, 640K of memory and 1MB of disk space are all that are necessary for efficient operation. Installation takes a few minutes. The program is straightforward and easy to operate. Rather than the months it typically takes to figure out the complicated software packages sold today, you'll be using FibNodes effectively within an hour. To fully implement all of the program's features will take you some time however, depending on how often you trade.

Your choice of a graphics program to use with FibNodes is important. If you're trading instruments like a five minute S&P, your graphics software should have the capability to resize charts both vertically and horizontally with a minimum of key strokes. Changing Time Frames should be an effortless event. Cursor access to critical chart locations must be quick and easy. Your choice of graphics software should also include the capability to *properly* calculate continuous contracts. See Appendix I.

Many of my clients want to see an automatic version of FibNodes and I am in discussions with a number of high quality graphics software vendors to explore this possibility. What I have in mind with an automatic version is a more efficient method for *the user* to insert data into FibNodes. I do not anticipate nor do I expect to be a part of any attempt to automatically generate FibNodes support and resistance levels directly from a chart, *absent of thought*. I've been down this road with the neural network crowd and the results were not good. Although such a program could easily be sold for a lot of money, its output, based on what I have seen, would be of minimal value. If this could ever be done and such software was easily attainable, the functionality and implementation of the concept could be in question. See CHAPTER 1.

COST:

The price of FibNodes has been reduced from $795 to $295, while its features and functionality have been increased. The cost of the software does not justify fooling around with spread sheet programs in an attempt to duplicate what has taken me 10 years to achieve. Many of FibNodes' advantages are subtle. They were developed for fast-paced trading and cannot be appreciated until you have worked with the concept and implemented the strategies over a period of time.

FibNodes™ FEATURES:

The power of the program comes from its ease of use, unique presentation, and organizational capabilities. Below are some of its features:

- HOT KEYS: virtually all FIBNODE FUNCTIONS can be implemented by a *single key input*. There are no complicated command lines and no fussing around with the mouse.

- PAGING CAPABILITY: You can efficiently access previously created FIBNODE files by an alphabetized pull down menu or by simply hitting the "+" or the "-" key.

- RATIO SELECTION: FibNodes defaults to the ratios I use, but you can easily *select and maintain* virtually any ratios you find useful. This can also be used as an excellent research tool.

- DUPLICATE A FILE: This feature allows you to automatically create the *same file* with *different* ratio inputs.

- TOGGLE KEY: By pressing "T" you will instantly see two additional numbered Fibnode locations at the .5 and .79 ratios.

- 32ND CONVERSION: The program recognizes bond, treasury notes, or muni bond files and automatically accepts and produces Nodes in 32nds notation.

- COLORED NODES: As a visual aid, Focus and Reaction Numbers are in white, while the date, time stamp, and ratio selection appear in blue. Support Nodes and Profit Objectives are shown in *Green*, while resistance Nodes and Profit Objectives are shown in *Red*.

- UPDATING FILES: You can change Focus Numbers by simply pressing "F". "A" will add while "D" will delete *multiple* Reaction numbers. You can also choose *where* Reaction numbers are to be placed within the series. .

- Plus many more features

For new developments and updates you can check our web site or call our office.
www.fibtrader.com • email coast@fibtrader.com • 941 346-3801 • Fax 941 346-3901

APPENDIX G

THE OSCILLATOR PREDICTOR™:

Below is an example of how the Oscillator Predictor study works in practice.

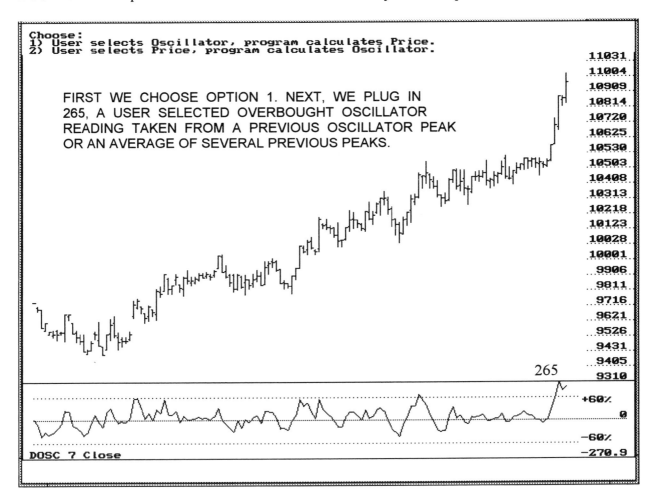

CHART AG-1

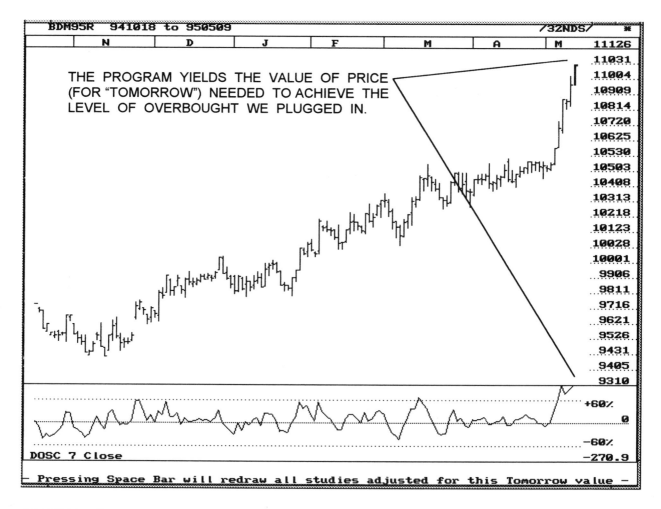

THE PROGRAM YIELDS THE VALUE OF PRICE (FOR "TOMORROW") NEEDED TO ACHIEVE THE LEVEL OF OVERBOUGHT WE PLUGGED IN.

CHART AG-2

Now we have a means of capturing profit at extremes in price but, there's a lot more we can do. We can use this tool to play "what-if" games. The same set of parametric equations will give us the level of the Detrended Oscillator for any given price, thereby allowing us to filter a trade as described in Strategy 2, CHAPTER 7. Here's how it would work in practice. You know the value of the DMA ahead of time, i.e. for tomorrow. You know price must exceed that value to get an entry signal. Plug the value of the DMA into the program, using Option 2, Chart A-G1 and you will know ahead of time, the level of Overbought and Oversold *corresponding to that price level.* Now you can make an informed decision, whether or not you wish to take that trade.

OSCILLATOR PREDICTOR™ AS APPLIED TO THE SHORT TERM S& P TRADE DESCRIBED IN CHAPTER 15.

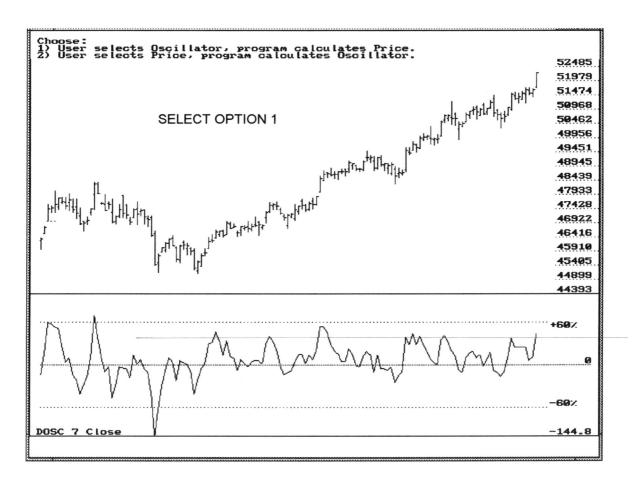

CHART AG-3

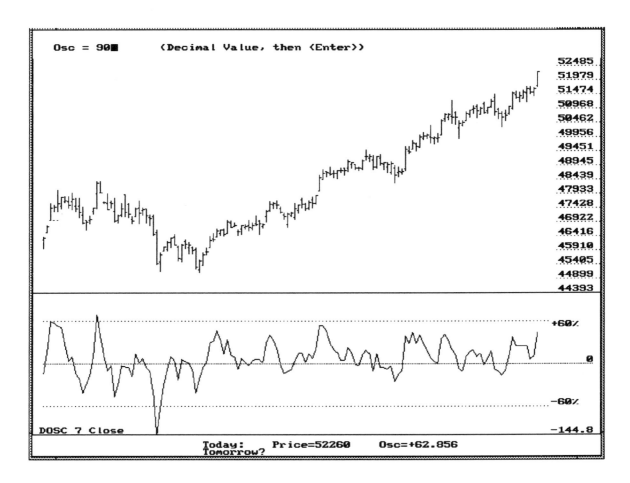

CHART A-G4

To achieve an Oscillator value of 90, we would have to reach a price of 52730 (tomorrow).

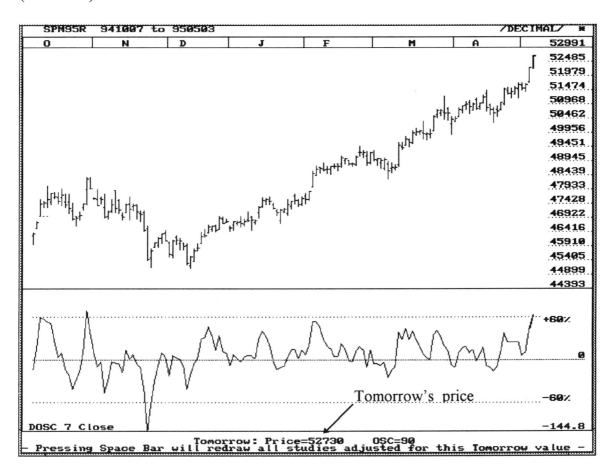

APPENDIX H

🗆🗆

SHORT TERM S&P TRADE FROM CHAPTER 15, TIME & SALES

This depiction of T&S is a more accurate representation of what actually occurred on the floor. It differs substantially from what I was viewing at the time of the trade.

SPM5 — CME S&P 500 Stock Index — Future							Time & Sales
5/ 4/95							
15:07 523.70	523.65	523.60	523.55	523.50	523.45	523.40	523.40
523.35	523.30	523.25					
15:08 523.20	523.15	523.10	523.15	523.20	523.30	523.25	523.20
523.15	523.10	523.20	523.15				
15:09 523.10	523.05	523.00	522.95	522.90	522.85	522.80	522.75
522.80	522.75	522.70					
15:10 522.60	522.50	522.45	522.40	522.50	522.55	522.60	522.65
522.70	522.60	522.70	522.75	522.80	522.75	522.70	
15:11 522.60	522.50	522.40	522.35	522.30	522.20	522.10	522.00
521.90	521.70	521.60	521.50				
15:12 521.40	521.30	521.20	521.10	521.00	521.10	521.20	521.30
521.50	521.60	521.70	521.80	521.90			
15:13 522.00	521.70	521.50	521.60	521.70	521.80	521.90	522.00
521.95							
15:14 521.90	522.00	522.00	522.10	522.20	522.30	522.50	522.40
522.00	522.20	522.30					
15:15 522.40	522.20	522.00	522.10	522.15	522.20	522.10	522.00
521.90	522.00						
15:16 522.10	522.20	522.10	522.00	521.90	522.00	522.20	522.30
15:17 522.40	522.30	522.20	522.10	522.00	522.10	522.00	522.20
522.00	521.90						
15:18 521.80	521.75	521.70	521.60	521.70	521.80	521.90	521.80
521.70							
15:19 521.60	521.50	521.40	521.45	521.30	521.20	521.10	521.00
521.10	521.20	521.20					

APPENDIX I

CONTINUOUS CONTRACT CREATION:

There are an almost unlimited means of creating continuous contracts from individual Futures contracts. The following method works best when using the Fibonacci Analysis techniques taught in this book

Use all available contract months, regardless of volume or open interest considerations. Include the current months data right up to the last trading day of the contract. Do not adjust for any gap that may occur when you append the next months contract to the last day of the current contract. Friday should be considered to be the close for the week, while the end of the month, regardless of which day it falls on, is considered to be the close of that month.

APPENDIX J

ᴄᴄ

COAST INVESTMENT SOFTWARE, INC. products and services:.

6907 Midnight Pass, Sarasota, FL 34242   • 941 346-3801   • Fax 941 346-3901

• e-mail coast@fibtrader.com   • www.fibtrader.com

TEACHING AND TRAINING:

Joe DiNapoli's, **"FIBONACCI, MONEY MANAGEMENT, AND TREND ANALYSIS, in home trading course"** This course consists of eight 90 minute audio tapes, an accompanying 100 page Instruction Manual, a high quality, wide span, Proportional Divider and an Applications Manual which describes in great detail how to properly use the divider. This is a recorded, 2-day seminar detailing a wide variety of topics *essential for a trader's success*...$375.00
without the Proportional Divider and an Applications Manual.............................$275.00

The **PROPORTIONAL DIVIDER & APPLICATIONS MANUAL:** The Proportional Divider is a high quality architectural tool, manufactured and engineered in Germany. It is light weight and easy to handle. With this precision instrument, a trader can graphically locate important Fibonacci ratios, Confluence, and Lineage Markings. It is generally considered to be an *indispensable trading tool,* even if you have the FibNodes™ software. The Applications Manual is a detailed, 60 page, professionally illustrated, instruction manual describing the manipulation of the Proportional Divider, as well as the theory behind the use of this device...$129.00

PRIVATE SEMINARS: Two day, intensive tutorials are held in the trading room during trading hours. Attendance is limited, prerequisites apply, and availability is very limited. Contact the office for price and scheduling information.

TRADING SOFTWARE:

FibNodes™: A *unique* Fibonacci Retracement and Objective Calculator, designed specifically for hectic, high-pressure, intraday trading, as well as position trading where high accuracy stop placement and targeted "profit objectives" are a must! The Fibonacci levels are *calculated* and *presented* in a manner consistent with Joe DiNapoli's approach as detailed in *DiNapoli Levels*...$295.00

THE CIS TRADING PACKAGE: A complete, affordable computer software graphics trading package. It competently provides end-of-day analysis and has the proprietary "Oscillator Predictor™" study, as well as other studies Mr. DiNapoli recommends for competent trend analysis and Directional trading techniques...........................$295.00

BIBLIOGRAPHY:

Gerald Appel *The Moving Average Convergence-Divergence Trading Method* (Signalert Corp., 150 Great Neck Road, Suite 301, Great Neck, New York 11021)

Jacob Bernstein, *Short Term Trading in Commodity Futures*, (Probus Publishing Company, 1987, ISBN 0-917253-66-3),(MBH Commodities, 60 Revere Dr., #888, Northbrook, IL 60062, 800 678-5253)

Robert Edwards and John Magee, *Technical Analysis of Stock Trends*

Larry Ehrhart, 3700 North Lake Shore Drive, Suite 7-09, Chicago, IL 60613, 312 871-4687, 312 789-7434 ●(Volume Studies)

J.K. Hutson, "Filter Price Data: Moving Averages vs. Exponential Moving Averages" *Technical Analysis of Stocks & Commodities* magazine, May/June 1984.

George Lane. Investment Educators, 719 S. Fourth Street, Watseka, IL 60970 800 962-9836, (815) 432-4334 ●(Stochastics)

John Murphy, *Technical Analysis of the Futures Markets*, (New York Institute of Finance, New York 1986)

Markets & Market Logic by Peter Steidlmayer and Kevin Koy, (The Porcupine Press, 1986, ISBN 0-941275-00-0, 401 S. LaSalle St., Suite 1101, Chicago, IL 60605)

J. Welles Wilder Jr., *New Concepts in Technical Trading Systems* (Trend Research, 1978),

Bill Williams, Ph.D., C.T.A., Profitunity Trading Group, Ltd. 2300 Pilgrim Estates Dr., Texas City, TX 77590-3750 409 945-8880, Fax 409 945-8887, e-mail ptg@phoenix.net,   www.profitunity.com ●(Judgmental Trading)

REFERENCE MATERIAL:

The following references appear in this section because I believe they have the *potential* to offer you significant value. Pick and choose among them, as some are quite advanced while others are beginner level. Certain references are more for interest than practical trading application, while other references help you hold on to your profits after the trading is over. Listings in the bibliography *have not* been duplicated in this section. Next to certain references I have indicated a particular area of expertise or perhaps some suggested reading. Other references have no particular mention because the reference could have a number of publications or the form of the material provided does not lend itself to a brief description. I have been in the business for a long time and doubtless I have forgotten to include many worthwhile publications and individuals. To those I offer my sincere apologies.

Thomas Aspray, Boardwatch, 117 W 15th Ave, P.O. Box 2141, Spokane, WA 99210, 509 838-0434, Fax 509 747-7801

Bill Bay, 1065 US 1 North, Ormond Beach, Fl 32174 •(Volume Studies)

Thomas A. Bierovic, Synergy Futures, 519 Riva Court, Wheaton, IL 60187, 630 682-3768, Fax 630 682-3915

John Bollinger, Bollinger Capital Management, Inc., P.O. Box 3358, Manhattan Beach, CA 90266, 310 798-8855, Fax 310 798-8858

Walter Bressert, P.O. Box 8268, 9440 Doubloon Drive, Vero Beach, FL 32963, 407 388-3330, Fax 407 388-3389 •(Time Cycles)

Constance M. Brown, *Aerodynamic Trading* (New Classics Library, 1995, ISBN 0-932750-42-7, P.O. Box 1618, Gainesville, GA 30503) (Aerodynamic Investments Inc., 770 533-9161, Fax 770 536-1337 , e-mail CBspz&ibm.net, www.aeroinvest.com)

Bob Buran, Bob Buran Investment Vision, 8175 S Virginia Street, S850-359, Reno, NV 89511, 702 853-8667

Andrew E. Cardwell, Cardwell Financial Group, Inc., P.O. Box 1369, Woodstock, GA 30188 •(RSI, Divergence Techniques)

Michael Chalik, Universal Technical Systems, 6503 N. Military Trail, Suite 905, Boca Raton, FL 33496, 800 315-3893, Fax 561 989-9131, e-mail wetradeall@aol.com, www.tradefutures.com •(Non-judgmental Trading)

Laurence A. Connors and Linda Bradford Raschke *Street Smarts* (M. Gordon Publishing Group, Malibu, CA, 1995, ISBN 0-9650461-0-9), www.mrci.com/lbr/

Michael Gur Dillon, Symmetry Wave Theory, 1705 14th St., Suite 277, Boulder, CO 80302, 303 449-4601 •(Non-judgmental Trading)

Edward Dobson, *Understanding Fibonacci Numbers,* Traders Press, Inc. P.O. Box 6206, Greenville, SC 29606, 800 927-8222, Fax(803) 298-0221

Mark Douglas, Trading Behavior Dynamics, 195 N. Harbor Drive, Suite 1603, Chicago, IL 60601 312 938-1441, Fax 312 856-2184 •(Psychology)

Dr. Alexander Elder, The Russian Exchange, 157 West 57th Street, Suite 1103, New York, NY 10019, 212 962-6894, 718 639-8889

Peter Eliades, cyclese@earthlink.net •(Cycles)

Tucker J. Emmett, *Fibonacci Cycles and Commodity Price Behavior* (Tucker Emmett, Stotler & Company, 30 South Wacker Drive, Chicago, IL 60606, 312 930-1450)

Rober Fischer, *Fibonacci Applications and Strategies for Traders,* (John Wiley & Sons, Inc. 1993, ISBN 0-471-58520-3), *The Golden Section Compass*

Nelson Freeburg, Formula Research, 4745 Poplar Ave., Suite 307, Memphis, TN 38117, 901 767-1956, 800 720-1080, Fax 901 458-0066 •(Non-judgmental Trading)

William R. Gallacher, *Winner Take All* (Probus Publishing Company, 1994, ISBN 1-55738-533-5)

Joseph and Francis Gies, *Leonard of Pisa and the Mathematics of the Middle Ages*

Sunny Harris, *Trading 101 - How to Trade Like a Pro* (John Wiley & Sons, 1996)
Sunny Harris & Assoc., Inc.,2075 Corte del Nogal, Suite C, Carlsbad, CA 92009-1414, 888 68-Sunny, 760 930-1050, Fax 760 930-1055, www.moneymentor.com

Cynthia Kase, Kase & Co., 1000 Eubank NE, Suite.C, Albuquerque, NM 87112, 505 237-1600

P.J. Kaufman, *The New Commodity Trading Systems & Methods* (John Wiley & Sons, 1987), Maple Hill Farm, P.O. Box 7, Scotch Hollow Rd., Wells River, VT 05081, 802 429-2121

Robert Krausz & Jeanne Long, Fibonacci Trader Corp., 757 SE 17th Street, Suite 272, Fort Lauderdale, FL 33316, 512 842-1166, Fax 954 566-2427, e-mail fibbo@safari.net

Joe Krutsinger, Robbins Trading Company, Presidents Plaza, 7th Floor, South Tower, 8700 W. Bryn Mawr, Chicago, IL 60631-3507, 312 714-9000.

Charles Le Beau & David W. Lucas *Technical Traders Guide To Computer Analysis of the Futures Markets* (Business One Irwin, Illinois, 1992)

Lou Mendelsohn, 25941 Apple Blossom Lane, Wesley Chapel, FL 33544, 800 732-5407, 813 973-0496, Fax 813 973-2700, lm@profittaker,com

S. Edward Moore, *Rhythm of the Markets,* 8000 River Road, Suite 11C, N.Bergen, NJ 07047, 800 686-0833, 201 861-0993, Fax 201 295-8664, rhythmofthemarkets.com

Glenn Neely, Elliott Wave Institute, 1278 Glenneyre, Laguna Beach, CA 92651, 800 636-9283, Fax 714 493-9149

Larry Pesavento, 4625 E. Camino Rosa, Tucson, AZ 85718, 520 529-0469, Fax 520 529-0491, www.tradingtutor.com,•(Fibonacci Analysis)

Charles Plank, Pi Inc., 23130 Hartland Street, Canoga Park, CA 91307 ●(Fibonacci Analysis)

Robert Prechter, New Classics Library, P.O. Box 1618, Gainesville, GA 30503, 405 536-0309 ●(Elliott Wave)

Ted Tesser, Waterside Financial Services, 1035 Spanish River Road, #106, Boca Raton, FL 33492, 407 989-0642 ●(Tax Consulting for Traders)

Teweles, Harlow, & Stone, *The Commodity Futures Game,* (McGraw-Hill)

Dr. Van K Tharp, IITM Inc. 8308 Belgium Street, Raleigh, NC 27606, 919 233-8855, Fax 919 362-6020 ●(Psychology)

Ralph Vince, *The New Money Management* (John Wiley & Sons, 1995) www.technalink.com/rv.shtml ●(Money and Portfolio Management)

Larry Williams, Commodity Timing, 140 Marine View, Suite 204, Solana Beach, CA 92075, 619 756-0421●(Non-judgmental Trading)

SPECIAL MENTION TO:

Aspen Research Group, Ltd.
710 Cooper Avenue, Suite 300
P.O. Box 1370
Glenwood Springs, CO 81602
800 359-1121

Neal Hughes
11121 NE 97th Street
Kirkland, WA 98033
425 822-5210
neal@halcyon.com
● Internet Services

Elyce Picciotti, Ltd.
613 North St. Patrick Street
New Orleans, LA 70119
504 488-3651 Fax 504 486-6187
e-mail elyce@bellsouth.net

Steven E. Roehl
326 Pontevedra Lane
Niceville, FL 32578
850 729-7522
Fax 850 729-2441
e-mail roehl@cybertron.com

ABOUT THE AUTHOR

Joe DiNapoli is a veteran trader with over 25 years of solid market trading experience. He is also a dogged and thorough researcher, an internationally recognized lecturer, and a widely acclaimed author.

Joe's formal education was in electrical engineering and economics. His informal education was in "the Bunker", an aptly named trading room, packed full of computers and communications equipment, where most of Joe's early research began.

Exhaustive investigations into Displaced Moving Averages, his creation of the proprietary Oscillator Predictor™, and in particular, Joe's practical and unique method of applying FIBONACCI ratios to the price axis, makes him one of today's most sought after experts.

As registered C.T.A. for over 10 years, Joe has taught his techniques in the major financial capitals of Europe and Asia, as well as in the United States. In 1996 alone Mr. DiNapoli held workshops to capacity audiences in 23 financial centers around the globe. His contributions have appeared in a wide variety of trading books and technical analysis publications.

When Chuck LeBeau asked his readers for names of successful traders they most wanted interviewed, Joe DiNapoli came up more often than any other. Likewise the "Atlanta Constitution" cited Joe's work by referring to the "magical power" of Fibonacci ratios in the market place. Joe has used this magic time and again on national TV to make both startling and uncannily accurate market predictions, particularly in stock market indexes and interest rate futures.

As president of Coast Investment Software, Inc., located on Siesta Key in Sarasota, Florida, Joe continues to develop and deploy "high accuracy" trading methods, using a combination of leading and lagging indicators in unique and innovative ways. He conducts a limited number of private tutorials each year at his trading room and he also makes his trading approach available to others via software and Trading Course materials.